Bernd Philippen

Transfer path analysis based on
in-situ measurements for automotive applications

Logos Verlag Berlin GmbH

λογος

Aachener Beiträge zur Technischen Akustik

Editor:
Prof. Dr. rer. nat. Michael Vorländer
Institute of Technical Acoustics
RWTH Aachen University
52056 Aachen
www.akustik.rwth-aachen.de

Bibliographic information published by the Deutsche Nationalbibliothek

The Deutsche Nationalbibliothek lists this publication in the Deutsche Nationalbibliografie; detailed bibliographic data are available in the Internet at http://dnb.d-nb.de .

D 82 (Diss. RWTH Aachen University, 2016)

ISBN 978-3-8325-4435-5
ISSN 1866-3052
Vol. 27

Logos Verlag Berlin GmbH
Comeniushof, Gubener Str. 47,
D-10243 Berlin
Tel.:   +49 (0)30 / 42 85 10 90
Fax: +49 (0)30 / 42 85 10 92
http://www.logos-verlag.de

Transfer path analysis based on in-situ measurements for automotive applications

Von der Fakultät für Elektrotechnik und Informationstechnik
der Rheinisch-Westfälischen Technischen Hochschule Aachen
zur Erlangung des akademischen Grades eines
Doktors der Ingenieurwissenschaften
genehmigte Dissertation

vorgelegt von

Diplom-Ingenieur
Bernd Philippen
aus Heinsberg

Berichter:
Universitätsprofessor Dr. rer. nat. Michael Vorländer
Universitätsprofessor Dr.-Ing. Jens-Rainer Ohm

Tag der mündlichen Prüfung: 12. Mai 2016

Acknowledgements

It is a pleasure for me to thank those who made this thesis possible. I would like to thank my supervisors Prof. Dr. Vorländer and Prof. Dr. Ohm for supporting me. I discovered my interest in acoustics and signal processing in their lectures. I cannot imagine a better support.

I would like to show my deepest gratitude to Prof. Dr. Genuit. He gave me the possibility for doing my Ph.D. during my work at HEAD acoustics. His psychoacoustics lecture is the reason why vehicle acoustics is my profession. I thank him for promoting my career and trusting in my abilities. This thesis would not have been possible without Prof. Dr. Sottek. I thank him deeply for encouraging my research and for many fruitful discussions. He motivated me to do my best and I have learned a lot from him. I am indebted to all of my colleagues at HEAD acoustics for their support. They made it a good time and encouraged me to keep on going.

I would also like to thank my committee members Prof. Monti and Prof. Dr. Negra. They made my doctoral examination despite my excitement a pleasant time which I will gladly remember. I am very grateful to Densua Mumford for proofreading my thesis.

A special thanks goes to my family. Words cannot express how grateful I am to my parents, Waltraud and Josef. Without them I would not have achieved all this. I owe my deepest gratitude to my grandmother Mia who always believes in me. She never gives up.

Contents

Chapter 1

Introduction

Today, sound quality is becoming increasingly important for products and machines. In general, when a new technology is introduced, first of all, pure functionality will be in the foreground. Acoustic aspects may not be relevant at first. But during further developments ergonomics and ecology have grown more and more in importance. Finally, the customer's needs must be fulfilled in order to achieve a successful product. In this context sound quality is important. It cannot be disregarded because hearing is an important human sense.

Achieving good sound quality is about more than just reducing sound pressure levels. Sometimes the influence of a sound can even be subtle and subliminal. However, the product's sound can have a major influence on the product quality perceived by the listener [1]. Also important are emotions, which are linked to the acoustic experience. Customers' emotions have an increasing influence on buying decisions and long-term satisfaction, especially if the technical facts of competing products are very similar. Additionally, whenever humans are involved sound quality has a more or less strong influence on the acceptance of any kind of machine or vehicle. This is valid for both household appliances such as washing machines and luxury cars.

In this context the modern automobile is a typical example. Starting in 1886 with the first vehicle with a combustion engine [2], the automobile and the engine have been enhanced step by step for 130 years. At first, technical aspects such as developing faster ways to travel with higher range and more power compared to a horse-drawn carriage were the main points of fascination. Today, besides legal acoustic requirements, brand sound and sound quality have a big influence on the purchase decision [3], because the technical standard is quite similar among competing cars. Of course, for luxury or upper class cars, where money is not really an issue, this is more appropriate then for the sub-compact class. Driving experience is composed of both the driving dynamics and the acoustics of a vehicle [4].

1.1 Sound sources in a vehicle

Vehicle acoustics deals with mechanisms of sound generation and propagation in the vehicle interior and exterior. Its goal is to develop countermeasures to manipulate the sound, yielding an effective noise reduction or improvement. Countermeasures include a suitable constructional design of car components, including the noise sources and the selection of materials, such that they lead to proper sound absorption and insulation [5].

Figure 1.1: *Many sound sources contribute to the vehicle interior noise.*

There are many sound sources in a vehicle, which contribute to the interior noise as it is shown in Figure 1.1 [6]. Powertrain, tire-road contact and wind flow are the major sources of the driving noise [5].

1.1.1 Powertrain

In the case of a combustion engine the powertrain noise consists of contributions from engine, gear box, intake and exhaust system. Gas and mass forces, as well as gas exchange, produce noise depending on engine load and speed. In an electric vehicle there is an electric motor, an inverter and a battery cooling system, which generate noise. The powertrain noise should reflect the power of the engine and underline the positioning of the car on the market. A sports car has different requirements to a luxury sedan or a price-driven subcompact. For example, the typical character of a sports car's sound can be achieved by a thundering exhaust noise and an aggressive intake noise.

1.1.2 Tire-road contact

The tire-road noise generated by the contact between rolling tires and road surface has an influence on the driving comfort, in particular while traveling with a constant moderate speed and low engine load. It depends on road surface, material properties of the rubber and tread pattern. A very periodic tread pattern would lead to 'singing' tires with tonal noise components. The tire-road interior noise depends among other things on cabin insulation and suspension mounts, of which the design requires a balance between vehicle dynamics and acoustic comfort. Stiff mounts give good handling characteristics and poor acoustic comfort and soft mounts poor handling characteristics and good acoustic comfort.

1.1.3 Other sound sources

The wind flow induced noise increases with rising speed. It is a random noise, but side mirrors or the antenna can produce unwanted whistling noise, if they are not well designed. Driving at high speed can be annoying, if the wind noise is too loud.

In the last decades the powertrain noise has been reduced continuously, so that the noise of auxiliary devices is less and less masked. In the case of an electric vehicle there is almost no masking. This has the effect that the noise of auxiliary devices like oil pumps or air conditioning compressors become noticeable. Their noise can be disturbing or irritating. The abrupt noise of starting such a device, which is not related to the direct action of the driver, can be misinterpreted as failure. Squeaking brakes or grinding noises reduce the perceived reliability of the brakes, and these noises are thoroughly disturbing. Many electric motors are installed in a modern car, which all produce more or less noise. Power windows, electrical seat adjustment or the roof of a convertible should sound powerful, which does not necessarily mean loud, and durable without leaving the impression that each use could be the last.

1.1.4 Relevance and meaning of vehicle sounds

A vehicle sound is not only a by-product of a mechanical system under operation, but it can have a meaning and relevance. The sound of a closing door seems to be unimportant at first glance, but the right sound communicates attributes like robustness and safety. A wrong sound implies properties like cheap or unreliable. Opening and closing the door is the first contact with a new car in the show room. It gives a first impression about the product and its quality. First impressions count. Hence, car manufacturers attach increasing importance to designing the door sound [7] (Figure 1.2). A potential buyer judges with his visual, tactile and aural sense based on his experience, knowledge and attitude. Noise and vibrations are closely linked to each other. For example, the perceived loudness is influenced by whole-body vibrations [8]. Poor sound or vibration does not necessarily mean that the real quality is poor, too. The essential point is what the costumer perceives and believes. In his or her experience poor quality is often combined with poor sound, so he or she tends to generalize it to all situations. But on the other hand, if product sound is well designed, it can be actively used to transport a positive message.

Figure 1.2: *Car manufacturers attach increasing importance to door sound design [7]. Picture is taken from [9].*

In general, vehicle sounds can be welcome or undesired. For example, the engine noise gives feedback about the current driving condition, like speed or load. The driver of a sports car wants to hear and feel the power under the hood. Feedback is important because the driver would like to have information about how something is moving or operating. The sound of the direction indicator is helpful to check whether it is switched off or on. A sound can also indicate an upcoming failure, e.g., a squeaking V-belt. It is essential that the normal operating sounds of a vehicle do not sound like a defect. Otherwise the customer will go uselessly to the dealer's garage and will complain about a sound, which cannot be changed. Then the quality perception suffers and the customer's satisfaction decreases. On the other hand, squeaks and rattles are annoying. Such noises would be intolerable for a premium sector passenger car. Overall, the vehicle sound should be suited to the vehicle class and character according to the costumer's expectations [10].

1.2 Sound engineering methods

The meaning of vehicle sounds shows that nothing should be left to chance. It is clear that in the development process of a vehicle much effort is invested in acoustics. The acoustic engineers deal often with disturbing or annoying sounds. In a complex scenario, like a car, finding the cause of an unwanted noise and improving it is a challenging task. A clear, structured approach is necessary.

1.2.1 Interior noise measurements

The first thing that can be done is a recording of the interior noise while driving on a test track or during operation on a test rig. It is preferable to use a binaural recording system, like an artificial head [11], shown in Figure 1.3, or a binaural headset, because the acoustic impression, like auditory spaciousness, is preserved. A recording of the interior noise allows for the documentation and quantification of acoustic issues. Modifications or different solutions can be compared. Discovering the appearance of unwanted noise is necessary but not sufficient. This is only the first step. In the second more challenging step, the actual source or the cause of this noise needs to be identified. The question is how disturbing noise patterns get to the driver's ears. Then the next task is finding suitable countermeasures that will mitigate the offending issues.

Figure 1.3: *Artificial head - a binaural recording system*

1.2.2 Decoupling method

A simple but limited approach is the decoupling method [12], which was performed in the early 80s. Components are decoupled one after another and in each case the resulting noise is analyzed by comparing it to the sound of the initial situation. The relevance of each component is evaluated by how the noise changes when it is removed. This method is time consuming and requires mechanical modifications. It cannot be guaranteed that decoupling a component has no influence on the rest of the system. Nevertheless, the decoupling method is still used today to verify results gained from other more advanced methods [13].

1.2.3 Excitation signal analysis

Another approach, requiring more measurement equipment, is having a closer look at the sound sources. During operation the airborne sound radiation can be recorded next to each sound generating component of interest with a microphone. Structural vibrations can be measured with acceleration sensors (Figure 1.4). A relevant component can be identified, with some limitations, by analyzing the recorded signals and their relevance to the interior noise. If all recorded signals are highly correlated, the most dominant part cannot be found. All signals are more or less the same. But a component can be clearly identified if only one source signal contains a disturbing sound pattern like rattling. Excitation signal measurements help to perform an A-B comparison of different solutions. It can be used to document or to proof the effectiveness of modifications.

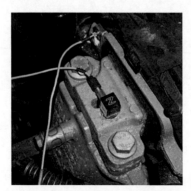

Figure 1.4: *A triaxial accelerometer needs only a small amount of space. It has, e.g., a size of 14x20x14 mm and can be easily applied on a structure using a mounting stud or adhesive [14].*

There is no need for mechanical modifications and multiple positions can be measured with a multi-channel measurement system at the same time. This method allows for only a qualitative statement about the contribution of the sources to the interior noise. The question of how a disturbing sound pattern gets to the interior cannot be answered. The transmission from the sources to a receiver, the point of interest, such as the driver's ears, is not covered.

1.3 Transfer path analysis and synthesis

The transfer path analysis and synthesis (TPA/TPS) approach combines receiver, source signals and transmission (Figure 1.5). It is a commonly used tool for sound quality troubleshooting and sound design, which is often used in the automotive industry to find suitable countermeasures for improving noise issues [15]. This approach identifies the sources and dominant transfer paths of unwanted noise components. In a next step this information can then be used for a specific manipulation in terms of sound design. For example, the perceived sportiness of a vehicle can be increased by emphasizing those transfer paths, which causes a roughness in the engine noise, so that the intensified roughness leads to a more sportive sound. In an early phase of the development process TPA/TPS can be used to predict the sound [16]. How will component A or B sound in vehicle X or Y? Test-bench measurements of an engine can be combined with vehicle data, too [17]. Even simulation data can be included [18].

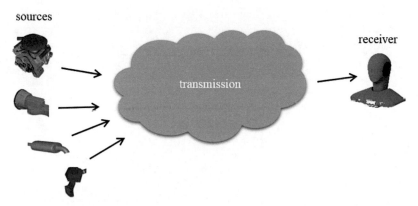

Figure 1.5: *Transfer path analysis is based on a source-transmission-receiver model.*

1.3.1 Source-transmission-receiver model

In general, TPA is based on a source-transmission-receiver model (Figure 1.6). Sound generating sources are transmitting to a given receiver position, which could be, for example, the driver's or operator's ears. In the automotive case the receivers are often the ears of the passengers and the driver especially. The signals can be measured with a microphone or an artificial head at the receiver position during operation of the vehicle. Of course, vibration receivers like seat rail or steering wheel vibrations are possible, too. Transfer path analysis tries to explain the measured receiver noise by a model of sources and their transfer paths.

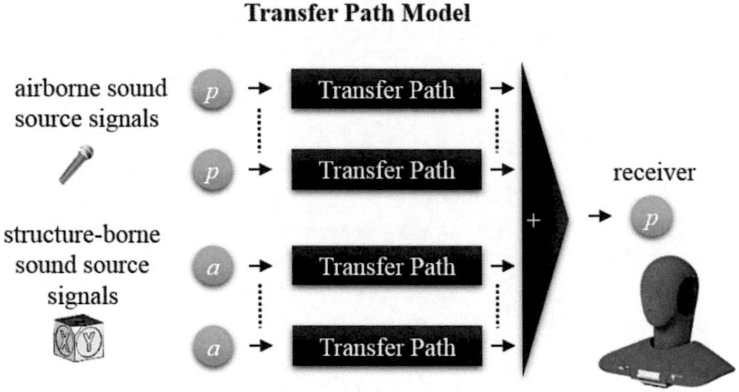

Transfer Path Model

Figure 1.6: *In a transfer path analysis model the receiver is a sum of transfer paths of source signals.*

1.3.2 Transfer paths

A transfer path describes the transmission from a source to a receiver (Figure 1.6). The superposition of the contributions from all transfer paths is a prediction of the receiver. A transfer path is modeled by a source signal and transfer functions. A transfer function characterizes the transmission behavior of the system in the frequency domain. They are often called frequency response functions (FRF). A FRF contains the frequency dependent attenuation or amplification and the delay from the source to the receiver. A transfer path can also consist of several transfer functions in a sequence, e.g., to describe the mount, the structure and the radiation characteristics. Several TPA methods exist to determine the transfer functions of a source-transmission-receiver model. Which method should be used depends on the actual application and other constraints like effort, time and budget. For instance, a detailed engine TPA requires other methods than a quick and short tire-road noise analysis. In general, methods differ in complexity and expected quality of results. Another aspect is the available measurement equipment, e.g., type of sensors or number of channels, which can limit the applicable methods. The experience of the acoustic engineer helps to find the best suited method. He must make an appropriate choice of method in each individual case [19].

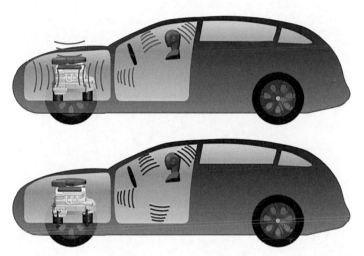

Figure 1.7: *Top: A source radiates sound into the ambient air, which reaches the receiver through different paths as airborne sound share. Bottom: If a source induces forces into a structure, the resulting vibrations are emitted as airborne-sound to the receiver. This part is called structure-borne sound to emphasize the relevance of the structure for this sound share.*

Transfer paths can be distinguished between airborne sound paths and structure-borne paths. A source radiates sound into the ambient air, which reaches the receiver through different paths as airborne sound share (Figure 1.7 top). A source is often connected to a structure, like an engine which is mounted on a car body. The operating source induces forces into the structure and the resulting vibrations are emitted as airborne sound to the receiver (Figure 1.7 bottom). This part is called structure-borne sound to emphasize the relevance of the structure for this sound share.

1.3.3 Input signals

The sound sources are described by signals measured during operation. These are also called input signals of the TPA model. The procedure is identical to excitation signal analysis. The airborne sound radiation of a source is recorded by a microphone in the near field. The distance between microphone and source is around 5 cm to 15 cm. If the source exceeds certain dimensions so that it cannot be assumed as a point source, several microphones are placed around the source. For an engine six microphones are typically used, one at each side [20], as it is shown in Figure 1.8.

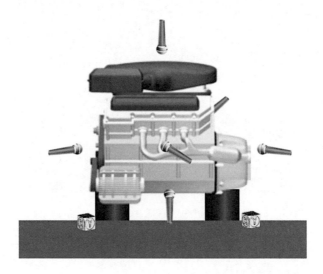

Figure 1.8: *The airborne radiation of an engine is typically measured using one microphone at each side.*

The reason for the structure-borne sound shares are forces which are induced into a structure. If these forces cannot be measured directly, acceleration signals are used as input signals of the TPA model as it is explained in the next chapter. All input signals must be recorded synchronously, otherwise artifacts like cancellation, exaggeration, and beat tones can occur. If a simultaneous data acquisition is not possible, the signals must be synchronized in a post-processing step [21].

1.3.4 Performing TPA

Input measurements

A TPA is usually performed in two steps. After setting up the test subject with sensors and the recording system, the first step is a measurement of the source and receiver signals under the operating conditions of interest.

The recording system should be portable with compact size and an own power supply. The type of the sensors and their positions depend on the chosen TPA method (see also next chapter). For example, in the case of an engine TPA, the car is running on a 4-wheel chassis dynamometer in an hemi-anechoic room [22]. An example is shown in Figure 1.9. The car is fixed on the test facility and the tires are put on four big rolls, which can be driven or braked. The operator has full control over gear, engine speed and load. A measurement can be reproduced exactly, and it is weather independent.

Figure 1.9: *The engine noise can be recorded on a 4-wheel chassis dynamometer in an hemi-anechoic room under fully manageable conditions.*

Frequency response function measurements

The frequency response function $H(f)$, also called transfer function, of a linear time-invariant system is the relation between the input signal $X(f)$ and the response $Y(f)$ at the system output (Figure 1.10).

$$H(f) = \frac{Y(f)}{X(f)} \tag{1.1}$$

Figure 1.10: *Transfer function $H(f)$ of a linear time-invariant system*

If the system output contains additive uncorrelated noise, as it is shown in Figure 1.11, the transfer function can be estimated using correlation techniques [23]. The cross spectrum of signal X related to signal Y is divided by the auto spectrum of X. The influence of the noise on the transfer function estimation is averaged out.

$$H(f) = \frac{S_{XY}(f)}{S_{XX}(f)} \tag{1.2}$$

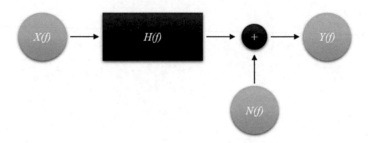

Figure 1.11: *Transfer function $H(f)$ of a linear time-invariant system with output noise*

The cross spectrum and auto spectrum can be estimated by averaging N repetitions [24] of simultaneous measured input and output signals.

$$H(f) = \frac{\frac{1}{N}\sum_{n=1}^{N} X_n^*(f) \cdot Y_n(f)}{\frac{1}{N}\sum_{n=1}^{N} X_n^*(f) \cdot X_n(f)} \tag{1.3}$$

In this thesis, the operator $\oslash$ is used to indicate that the transfer function is calculated based on correlation techniques.

$$H(f) = Y(f) \oslash X(f) \tag{1.4}$$

In the second step of TPA, the transfer functions, which are needed for the transfer path model, are determined in a test room optimized for acoustics (Figure 1.12). A hemi-anechoic room with sound-absorbing walls and ceiling is preferred. It should be a very quiet room, which is decoupled from the rest of the building, so that no noise or vibrations from the exterior disturbs the measurements. A lifting platform for vehicles is helpful in many situations.

Figure 1.12: *Transfer functions are determined in a test room optimized for acoustics.*

1.3.5 Transfer path synthesis

The process of combining the source signals with the transfer functions describing the transmission to the receiver location is called transfer path synthesis.

The contribution of each path to the receiver sound is calculated. The synthesis of all transfer paths is the prediction of the receiver sound, which can be compared to a simultaneously measured reference signal at the receiver position for validation purposes. A good accordance between prediction and measurement is a hint for a good model quality. Considerably more important are the sound shares of the different sources and individual transfer paths, which can be calculated by the transfer path synthesis, too.

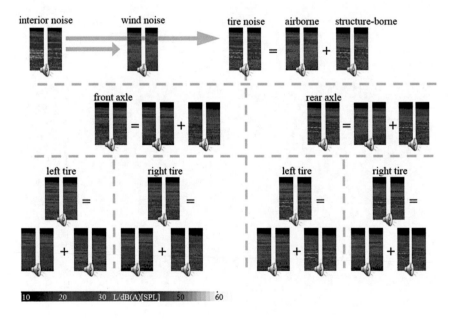

Figure 1.13: *A contribution analysis for tire-road noise splits the interior noise into its noise shares. The spectrograms of the time signals are shown to give a quick overview.*

1.3.6 Contribution analysis

In general, not all transfer paths contribute to the receiver noise to the same extent. The task of an acoustic engineer is to identify those paths, which have a noteworthy contribution, because only a modification of dominant paths gives sufficient potential for improvements.

A contribution analysis example is shown for tire-road noise in Figure 1.13. In a contribution analysis the cause of a disturbing sound pattern can be determined by evaluating each path or groups of paths. Dominant transfer paths are identified, which are primarily responsible for a disturbing sound pattern at the receiver position. They are also called critical transfer paths. The breakdown of the receiver noise into sources and transmission allows a root cause analysis [25]. It can be retrieved if the excitation or the transmission behavior or both are the reason for the offending issue. This gives additional information for possible improvements.

1.3.7 Frequency or time domain TPA

A distinctive feature of a TPA method is whether it is performed in the frequency or time domain [26]. In the first approach both the algorithm and the final result - the transfer path synthesis - are performed in the frequency domain. The frequency domain is a representation of a signal with respect to frequency, rather than time [27]. A time signal can be converted between time and frequency domain with the Fourier transform [28]. The time signal is transformed into a weighted sum of sine waves of different frequencies representing the frequency components. The frequency domain representation is called spectrum.

A frequency domain TPA allows for the analysis of which frequency components of the receiver signal are caused by which source and which transfer path. Knowledge of the frequency range helps in deriving conclusions about the underlying origin of the sound generating mechanism or gives hints about possible countermeasures. The transfer path synthesis is represented, for example, as averaged spectra or order levels vs. time in the case of a powertrain TPA. Orders are multiples of the rotating frequency; they link frequency and rotational speed. The evaluation is carried out versus rotation, because the rpm determines the operating point of the engine.

In the other case, one refers to a time domain TPA, if the final results are time signals. A time-domain approach is essential for analyzing transient sounds [29]. In the transfer path synthesis, the source signals are filtered in the time domain according to the transfer functions to synthesize the transmission to the receiver. Nevertheless, the transfer functions are still calculated in the frequency domain. But the difference is, that the user can listen to results at the end. This is often called auralization to emphasize the importance of a human as a listener. Averaged spectra and order or noise levels are, without question, helpful tools in the development process of a vehicle or a product, but only listening to TPA results allows a reliable assessment of the perceived sound quality. Thereby, auralization, that means the aurally adequate reproduction of time signals [30], plays a key role.

Customers associate consciously or subconsciously the quality of the whole product with its acoustics [31]. If you can make a what-if analysis audible, e.g., in a vehicle sound simulator [32], it is simpler and more intuitive to decide for or against an acoustic measure with the goal of improving the perceived quality. Another advantage of a time domain TPA method is that in a post-processing step any arbitrary analysis can be used. This covers averaged spectrum, order level vs. time but also a psychoacoustic analysis like loudness, roughness and sharpness.

1.4 Panel contribution analysis

Transfer path analysis and synthesis is a tool for identifying dominant sources and transfer paths. If a receiver is located inside a closed compartment, like a vehicle cabin, it is also relevant to know which panel has the strongest contribution to the vehicle interior noise and which partial surfaces are most productive for applying acoustic countermeasures. Answering these questions is beyond the definition of TPA used in this thesis.

For this purpose, a (binaural) panel contribution analysis must be performed [33]. A vehicle is subdivided into panels like roof, floor, cockpit and doors. The velocity measured close to the vibrating panels is combined with transfer functions to the receiver to evaluate the relevance of a panel. This method is not used only for passenger cars [34], but also for trains [35] and helicopters [36]. The results of a binaural panel contribution analysis can be visualized by coloring the panels corresponding to their relevance as it is shown in Figure 1.14. Furthermore, a distribution of sound contributions within each individual panel is available.

Binaural panel contribution analysis can be also combined with Binaural Transfer Path Analysis and Synthesis. This gives more a detailed information about the dominating panels for each transfer path [37].

Figure 1.14: *Binaural panel contribution analysis is a method to determine which panel has the strongest contribution regarding the disturbing noise.*

1.5 Objectives

Objectives of this thesis are in-situ transfer path analysis methods for automotive applications which cover the main sources of the driving noise: powertrain, tire-road contact and wind flow.

Different applications require different methods. The first part contains a TPA method for structure-borne sound induced via elastic elements with an improved prediction quality and which delivers more insight into the transmission behavior. Two constraints are of major importance: First, a good auralization quality, which implicates having time signals as final results. Auralization is a key feature for the assessment of perceived sound quality, but auralization artifacts can distract from the proper sound pattern and should be avoided. Methods and strategies are discussed to detect and reduce these artifacts in the transfer path synthesis. Second, mount characteristics are parameterized from in-situ measurements without the need for elaborate and often questionable test rig data considering the strong coupling of the receiver structure. Furthermore, a directional coupling within a mount can be modeled. A new approach is presented which combines the benefits of widely used inverse force identification method and the advantages of the mount attenuation function method. Mount characteristics, the effective mount transfer functions, are calculated in-situ from operational data considering the crosstalk on the car body. The effective mount transfer functions are an ideal starting point for estimating parameters of mount models. Physically motivated parameter models with only a few parameters are used to avoid overfitting and, therefore, meaningless values.

In the last few decades engine noise has been reduced continuously, so that the noise generated from rolling tires attracts more and more attention for sound improvement. With regard to the electrification of the powertrain, tire-road noise will become even more important for sound quality and NVH comfort because of lower masking by the electric motor noise. The second part of this thesis will present further improvements of tire-road noise analysis using on-road measurements. An approach for the auralization of tire-road interior noise under dynamic driving conditions without disturbing noise shares from engine and wind is proposed. In addition, the wind noise share can be synthesized. So far, the tire-road noise synthesis has been restricted to coast-down measurements with the engine switched off. With the new approach it is possible to auralize the tire-road noise during a run-up without disturbing engine noise shares to study the influence of different load and drive torque.

This work is a composition of contributions to conferences and journals. In chapter 2 the fundamentals of transfer path analysis and synthesis are explained. The advantages and drawbacks of different methods are discussed and an overview of the state of the technology is given. The focus is on structure-borne sound induced by a source which is elastically mounted on a structure.

Chapter 3 deals with parameterizing mount models based on in-situ measurements for engine TPA. The first part introduces a new approach called effective mount transfer function method, which considers the strong coupling of the receiver structure and takes the mounts into account using only in-situ measurements. Then mount models are parametrized to give a deeper physical insight into the transmission behavior. Examples and results are shown in the subsequent chapter 4.

The next chapter 5 is dedicated to methods for chassis TPA and tire-road noise analysis. A new approach is presented which allows for a synthesis of tire-road interior noise under dynamic driving conditions. Results and examples are given in the following chapter 6. Finally, conclusions are drawn and an outlook is given in the last chapter 7.

Chapter 2

Fundamentals of transfer path analysis

The focus of this thesis is on structure-borne sound induced by a source which is elastically mounted on a structure (Figure 2.1). This situation occurs in different fields of application. It could be an engine in a vehicle, like a car, a truck or a ship, as well as a compressor in a refrigerator or a machine tool mounted on its machine base.

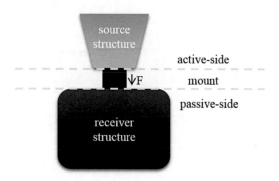

Figure 2.1: *A source structure (active side) induces forces into a receiver structure (passive side) via a mount.*

The fundamentals of transfer path analysis and synthesis for structure-borne sound are explained in this chapter. An overview of the state of technology is given by discussing advantages and drawbacks of different methods.

2.1 Structure-borne sound transfer path analysis

2.1.1 Applications for structure-borne TPA

In the automotive industry TPA is often used to find solutions for structure-borne noise issues, like a power steering pump transmitting from the hydraulic system to the car body structure [38], or idle boom noise caused by improper subframe mounts [13], or diesel knocking [39]. Another example is identifying major noise paths of axle gear noise, which is perceived as a tonal whine that can cause customer dissatisfaction [40]. Subsystem level targets can be quantified using the technique of transfer path analysis, too [41]. The targets for a subsystem are derived from an overall target according to its contribution to the overall noise. Experimental transfer path analysis is not only used for passenger vehicles but also for heavy duty trucks, e.g., for investigating cabin suspension [42].

The noise levels generated by domestic machinery products affect their commercial success. Refrigerators and washing machines are typical examples where a sound generating source is attached to a structure. If this is not well designed acoustics become an issue. Another example is that the noise level of a machine tool can affect the health and safety of working environments.

The designer of military ships needs to reduce the sound radiation of the propulsion system through the hull structures to minimize the chance of detection [43]. Transfer path analysis can help to optimize hydro-acoustic receiving antennas of submarines, so that their detection performance is less disturbed by noise generated by the submarine itself [44]. The comfort level on a cruise ship should not be affected by disturbing noises. Noise level prediction is important in order to comply with ship owner requirements [45]. In all these cases transfer path analysis is a helpful tool in the development process.

2.1.2 Substructuring

In many TPA models a substructuring is performed [46]. That means, the system is separated into source structure, which is called active side, and receiver structure, which is passive side. A mount can be used to decouple the source from the receiver structure leading to a substructuring into source, mount and receiver. Of course, a separation into more parts, like engine, mounts, mount brackets, subframe and car body, is possible with greater effort.

The source induces forces into the structure during operation. The result is a vibrating structure and the vibrations are radiated as structure-borne sound to the receiver (see Figure 1.7 bottom).

2.1.3 Mechanical impedance and mobility

If an ideal force source, concentrated at one point, acts on a structure, the resulting vibration at the contact point depends on the mechanical impedance of the structure. How much a structure resists motion when a harmonic force is applied, is given by the mechanical impedance. It is the ratio of the complex-valued amplitudes of the force and velocity at an interface for a given frequency [47].

$$Z(f) = \frac{F(f)}{v(f)} \tag{2.1}$$

Sometimes it is more appropriate to use the inverse of impedance, termed as mobility. The mobility is defined as the complex-valued ratio of the velocity across a structure element to the force through it in the frequency domain [48].

$$Y(f) = \frac{v(f)}{F(f)} \tag{2.2}$$

In contrast to velocity acceleration can be easily measured with widespread 1D or 3D accelerometers. That is the reason why the inertance, or sometimes called accelerance, is often used, which is the frequency response function (FRF) between force and acceleration.

$$I(f) = \frac{a(f)}{F(f)} \tag{2.3}$$

2.1.4 Electromechanical analogies

A mechanical system can be modeled as an interconnection of masses, springs and dampers. Representing it by an analogous electrical system is advantageous because the theory of analyzing complex electrical systems can also be used for the mechanical system. The mathematical behavior of the equivalent electrical system is identical to the mathematical behavior of the represented mechanical system.

With the help of electromechanical analogies mechanical quantities are transferred into electrical quantities and vice versa (table 2.1) [28]. Each element in the mechanical domain has a corresponding element in the electric domain with an analogous constitutive equation. In the following, the mobility analogy is used. Mechanical force corresponds to electric current and velocity to voltage. The topology of the mechanical system is preserved when transferred to the electrical domain, but mechanical impedance is not related to electrical impedance but to electrical conductivity.

Table 2.1: *Electromechanical analogies*

Impedance Analogy				Mobility Analogy
voltage U	$\leftrightarrow$	force F	$\leftrightarrow$	current I
current I	$\leftrightarrow$	velocity v	$\leftrightarrow$	voltage U
electr. impedance Z_{el}	$\leftrightarrow$	mech. impedance Z_m	$\leftrightarrow$	electr. conductivity Y_{el}
resistance R	$\leftrightarrow$	friction losses w	$\leftrightarrow$	conductivity $1/R$
inductivity L	$\leftrightarrow$	mass m	$\leftrightarrow$	capacity C
capacity C	$\leftrightarrow$	spring n	$\leftrightarrow$	inductivity L

2.1.5 Coupling between source and receiver

The coupling between source and receiver can be explained by circuit diagrams as shown in Figure 2.2 for a system with one degree of freedom. The source is characterized by its velocity and mobility. The load mobility which is connected to this source represents the structure. Force and velocity at the interface depend on the combination of the two mobilities and the source signal itself.

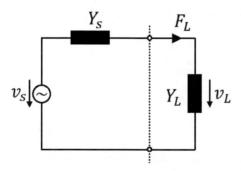

Figure 2.2: *The coupling between source and receiver can be explained by a circuit diagram.*

A source can be described by its mobility and the free velocity or the blocked force. The free velocity can be measured when no load impedance is attached, which corresponds to the case of an open circuit (left side of Figure 2.3). For this measurement, the source structure is mounted in a way that the source can move freely. The blocked force is measured with zero load mobility. In the electric domain this is a short circuit of the source as it is shown on the right side of Figure 2.3. During the blocked force measurement, the source is mounted on an infinitive stiff structure leading to no displacement at the interface. In practice, a sufficiently stiff structure compared to the source structure, e.g., a massive block of concrete, is used. If both the free velocity $v_{free}(f)$ and the blocked force $F(f)_{blocked}$ is known, the source mobility can be calculated.

$$Y_s(f) = \frac{v_{free}(f)}{F_{blocked(f)}} \tag{2.4}$$

Another possibility of determining the source properties is performing two different measurements using well-known load mobilities. Then the free velocity and the source mobility are calculated using electric circuit theory.

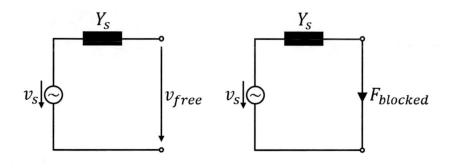

Figure 2.3: *A source can be described by its impedance and the free velocity or the blocked force. The free velocity can be measured when no load impedance is attached (open circuit). The blocked force can be measured with zero load mobility (short circuit).*

2.1.6 Structure-borne transfer path model

Forces act as load on a structure, which responds to the load with vibration. That means forces are the cause of velocity or acceleration. Therefore, these forces are the cause of structure-borne sound, too.

There are several methods for modeling structure-borne transfer paths and each has individual advantages and drawbacks. They are based on a load-response model: Forces are loads of a structure which respond with vibrations. The general transfer path model of Figure 1.6 can be extended by splitting a transfer path into two parts (Figure 2.4). In the first step of a structure-borne TPA model the operational forces are calculated. How this is being done in detail depends on the actual method. In the second step, acoustic transfer functions (ATF) describe the transmission to the receiver.

Transfer path model – structure-borne sound

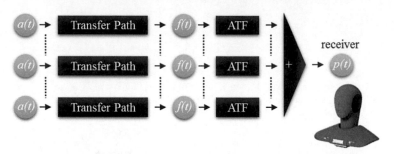

Figure 2.4: *The first step of a structure-borne TPA model is determining the operational forces. In the second step, the acoustic transfer functions (ATF) describe the transmission to the receiver.*

2.2 Auralization using acoustic transfer functions

The (vibro-) acoustic transfer function (ATF) is the frequency dependent relation between induced force and resulting sound pressure at the receiver position.

$$H^{ATF}(f) = \frac{p(f)}{F(f)} \tag{2.5}$$

For an auralization time signals are strictly necessary. That means a time domain transfer path synthesis is performed. Impulse responses are calculated from the transfer functions by means of Fourier transform in order to filter the signals of the input measurement in the time domain. After the operational forces have been determined according to one of the methods described in the following, they are filtered with the impulse responses of the ATF to calculate the resulting structure-borne sound and the sound shares of the transfer paths.

2.2.1 Reciprocal measurement of acoustic transfer functions

The ATFs can be measured directly with an impact hammer or reciprocal with a volume sound source in an anechoic measuring room [49, 50]. The principle of reciprocity states that the sound pressure at an observation point A produced by the vibration of a structure subject to the action of a point force at point B is the same as the velocity produced in the structure at point B, in the absence of the mechanical force, by a point source located at point A [47].

$$\left.\frac{p_B(f)}{F_A(f)}\right|_{Q_B(f)=0} = \left.\frac{v_A(f)}{Q_B(f)}\right|_{F_A(f)=0} \tag{2.6}$$

The system must be linear and time invariant. Another constraint is that the directional pattern of the volume velocity source must be equal to the directional characteristics of the receiver. In the case of a binaural recording system also a binaural volume velocity source must be used with the corresponding directional pattern [50] (Figure 2.5).

Figure 2.5: *The binaural volume velocity source has the same directional pattern like the artificial head recording system [50].*

2.2.2 Auralization based on active-side accelerations

If forces are determined indirectly from accelerations (as explained later), which are measured on the receiver structure, the structure-borne sound share auralization can be disturbed by additional sources that are not considered in the TPA model.

In the automotive example of an engine it could be the tire-road contact or auxiliary components which also contribute to the used acceleration signals. There is a remedy. In the case of high source mass and soft mounts the accelerations on the active-side are usually not affected by these components. Under these conditions an auralization based on active-side accelerations is preferable.

2.3 Direct force measurement

In a structure-borne TPA model forces are the starting point of the auralization. The most straightforward method to determine the operational forces is a direct measurement using force sensors.

The advantage of this direct approach is that no further excitations, calculations or signals are necessary. But in practice a direct force measurement is not possible with reasonable effort in a vehicle. It is time-consuming and a very elaborate task to place force transducers between the source and receiver structure. The available space, e.g. in the engine compartment, is usually limited so that there is not much space left for force sensors and adaptors [51]. Further challenging tasks include applying preload and calibrating the measurement cells.

Figure 2.6: *A triaxial force sensor is shown which is preloaded and precalibrated.*

In fact, there are preloaded and precalibrated force sensors available (Figure 2.6) [52], which are ready to use, but they need a lot of space (55x55x60 mm). In most automotive applications there is not enough installation space for these kinds of sensors. They could be potentially used on special test rigs. Another aspect is that the sensor must be capable of bearing the static load of the source. If adaptors must be crafted to allow for a mechanical connection between source and receiver structure, it has to be verified that this modification does not have an influence on the transmission in the considered frequency range.

If the forces can be measured directly, the structure-borne sound TPA model must be simple and it consists only of one transfer function, the ATF, per path. The related transfer path model is shown in Figure 2.7).

A direct force measurement in the vehicle is not common, but in [53] piezo force sensors are built into the strut bearing of a front axle. The damper struts are shortened to gain space for the force sensors. Here, the additional mass should not be relevant. In general, it is very difficult to ensure that the vibrating structure is not detuned due to a different force transmission geometry or added mass. In the mentioned example, another force sensor is built into the tie rod, but here the additional mass adulterates the measurement results.

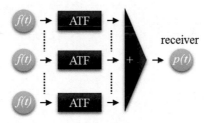

Figure 2.7: *The transfer path model of direct force measurement consists only of one ATF per path.*

2.4 Inverse force method based on mobility FRFs

Forces cannot only be measured directly with sensors, but they can also be determined indirectly from other quantities like acceleration measured during operation [54].

A widely used example for an indirect force determination (IFD) is the inverse force identification method, which is based on the inversion of measured mechanical mobility or inertance FRFs of the structure. The equations of indirect force determination can be formulated using force and velocity, as well as with force and acceleration. The inverse force identification method was already proposed in the early 80s to calculate operational forces induced into a car body. But at that time it was too elaborate to consider the coupling of more than three degrees of freedom at the same time. With rising computational power since the 90s, the method has been widely used for tire-road noise [55] as well as engine noise investigations [22].

This section contains a summary of the inverse force identification, further details are explained in [56, 57, 58]. The principle of this indirect method is that forces cause vibrations and, when it is the other way around, the operational vibrations measured with acceleration sensors can be used to determine the operational forces.

2.4.1 Inertance FRF measurements

[1] For the inverse force identification method (IFD), three-dimensional acceleration sensors are placed at the positions where the source induces forces into the structure. In the case of a car engine, they are applied next to the engine mounts on the car body as it is shown in Figure 2.8. At these positions, the structure is successively struck by an impact hammer in all spatial directions measuring the induced force and the resulting accelerations (Figure 2.9).

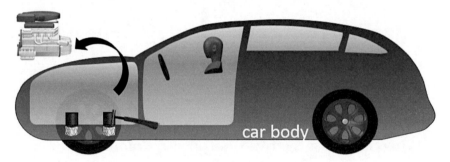

Figure 2.8: *Three-dimensional acceleration sensors are placed at the positions where the source (engine) induces forces into the structure (car body). At these positions, the structure is successively struck by an impact hammer in all spatial directions measuring the induced force and the resulting accelerations. During the impact measurements the source, e.g. the engine, is removed.*

During the impact measurements the engine is removed. From these measurements, the inertances, also referred to as accelerances, are calculated. The inertance $I_{k,l}(f)$ of a structure is defined as the transfer function [2] from the induced force F_k at the position k to the resulting acceleration a_l at the position l. To distinguish among the

---

[1]This part is a revised section taken from [59].

[2]The operator $\oslash$ is used to indicate that the transfer function is calculated based on correlation techniques.

different directions in space at each measured position, separate indices are used not only for each position but also for each direction of space.

$$I_{k,l}(f) = a_l(f) \oslash F_k(f) \tag{2.7}$$

The inertance FRFs are measured in a test room, which is at best a hemi-anechoic chamber with a solid floor. This fits best to the real life situation of a car driving on a road and it is a fully controllable test environment.

Instead of an impact hammer a small shaker with a force transducer can be used, too [60]. The reproducibility of a shaker compared to a manual hammer is higher, but a shaker is often difficult to mount. The coupling between shaker and structure may have an influence on the dynamic behavior of the structure which must be taken into account.

In general, the source structure needs to be removed to determine the inertances of the receiving structure only. Under certain circumstances a removal of a source structure is not necessary (refer to section 2.4.4 In-situ inverse force identification).

2.4.2 Derivation of inverse force identification method

Apparent mass transfer function

[3]At first, it is assumed that there is no coupling between the different positions or directions. That means a force at position k causes an acceleration only at $l = k$ and no acceleration at the other positions with $l \neq k$. In other words, each measured acceleration signal is only the result of a force acting at the same position and in the same direction. Vice versa, only the acceleration signal with the index k, which is recorded during operational conditions of the car or machine, needs to be considered to calculate the operational force F_k. The inertance is simply inverted to yield the apparent mass $H_k^{AMF}(f)$, which will be used to calculate the force indirectly from the acceleration:

$$H_k^{AMF}(f) = (a_k(f) \oslash F_k(f))^{-1} = \frac{1}{I_{k,k}(f)}. \tag{2.8}$$

In the time domain TPA, the operational force can be calculated by filtering the operational acceleration with the impulse response $h_k^{AMF}(t)$ of the apparent mass transfer function $H_k^{AMF}(f)$:

$$\tilde{F}_k^{AMF}(t) = a_k(t) * h_k^{AMF}(t). \tag{2.9}$$

---

[3]This part is a revised and extended section taken from [59].

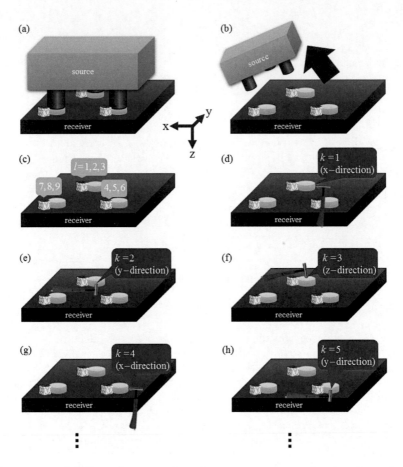

Figure 2.9: *(a) Three-dimensional acceleration sensors are placed at the positions where the source induces forces into the structure. (b) During the impact measurements the source is removed. (c) The acceleration signals are numbered with the index $l = 1..L$. (d)-(h) The structure is successively struck by an impact hammer in all spatial directions measuring the induced force with the index $k = 1..K$ and the resulting accelerations.*

Inertance matrix

However, as the car body is a strongly coupled system, a force acting at one position causes vibrations throughout the structure. Then, it is not sufficient to consider only the inertance for $l = k$ to calculate the force with the index k, but instead all combinations of impact and acceleration sensor positions must be evaluated. A frequency dependent inertance matrix $\mathbf{I}(f)$ with $k = 1 \cdots K$ and $l = 1 \cdots L$ can be set up:

$$\mathbf{I}(f) = \begin{pmatrix} I_{1,1}(f) & I_{1,2}(f) & \cdots & I_{1,K}(f) \\ I_{2,1}(f) & I_{2,2}(f) & \cdots & I_{2,K}(f) \\ \vdots & \vdots & \ddots & \cdots \\ I_{L,1}(f) & I_{L,2}(f) & \cdots & I_{L,K}(f) \end{pmatrix} \tag{2.10}$$

In the frequency domain, the operational accelerations are equal to the product of the inertance matrix and the operational force vector:

$$\begin{pmatrix} a_1(f) \\ a_2(f) \\ \vdots \\ a_L(f) \end{pmatrix} = \begin{pmatrix} I_{1,1}(f) & I_{1,2}(f) & \cdots & I_{1,K}(f) \\ I_{2,1}(f) & I_{2,2}(f) & \cdots & I_{2,K}(f) \\ \vdots & \vdots & \ddots & \cdots \\ I_{L,1}(f) & I_{L,2}(f) & \cdots & I_{L,K}(f) \end{pmatrix} \cdot \begin{pmatrix} F_1(f) \\ F_2(f) \\ \vdots \\ F_K(f) \end{pmatrix} \tag{2.11}$$

The coupling or crosstalk between spatial directions or different positions is described by the non-diagonal elements of the inertance matrix. All force signals can contribute to any acceleration signal. The above mentioned case without coupling is a special case leading to a diagonal matrix. Equation 2.11 can be expressed in a vector-matrix notation:

$$\mathbf{a}(f) = \mathbf{I}(f) \cdot \mathbf{F}(f). \tag{2.12}$$

The operational forces are calculated by multiplying the apparent mass matrix $\mathbf{M}(f)$ with the acceleration vector:

$$\mathbf{F}(f) = \mathbf{M}(f) \cdot \mathbf{a}(f). \tag{2.13}$$

This is a short notation for

$$\begin{pmatrix} F_1(f) \\ F_2(f) \\ \vdots \\ F_K(f) \end{pmatrix} = \begin{pmatrix} M_{1,1}(f) & M_{1,2}(f) & \cdots & M_{1,L}(f) \\ M_{2,1}(f) & M_{2,2}(f) & \cdots & M_{2,L}(f) \\ \vdots & \vdots & \ddots & \cdots \\ M_{K,1}(f) & M_{K,2}(f) & \cdots & M_{K,L}(f) \end{pmatrix} \cdot \begin{pmatrix} a_1(f) \\ a_2(f) \\ \vdots \\ a_L(f) \end{pmatrix} \tag{2.14}$$

Matrix inversion

If the number of acceleration signals L is equal to the number of force excitation positions K, the apparent mass matrix is the inverted inertance matrix. This approach is also called matrix inversion method to emphasize this calculation step of inverting matrices. The inversion is performed for every frequency f:

$$\mathbf{M}(f) = \mathbf{I}^{-1}(f) \text{ with } L = K. \tag{2.15}$$

The accuracy and numerical stability can be improved by using additional acceleration sensors applied on other positions on the structure [61]. At these indicator positions, impact measurements are not performed, but the related inertances are calculated. This leads to a matrix $\mathbf{I}(f)$ with $L > K$ describing an overdetermined system of equations which is solved using the pseudo-inverse (pinv) [62]:

$$\mathbf{M}(f) = \text{pinv}(\mathbf{I}(f)) \text{ with } L > K. \tag{2.16}$$

The inverse force identification method allows several model configurations. The most complex model considering the complete cross-coupling uses a dense inertance matrix with all combinations of impact and acceleration sensor positions and spatial directions.

If the coupling between some positions is so low, that it can be neglected, the corresponding elements in the inertance matrix can be set to zero yielding a sparse matrix. The measurement and calculation effort can be reduced, because these inertance functions are not needed. In this case it is better to ignore a low coupling than to record only random noise during impact testing which would result in wrong inertance FRFs having a bad influence on the matrix inversion. A 3x3 block diagonal matrix is used to consider the crosstalk only within each mount [63]. In the simplest case the assumption is made that the operational acceleration at one position is only caused by the force acting at the same position. Forces at other positions do not contribute to the considered acceleration. That means only the diagonal of each inertance matrix is inverted (diagonal method) which is equal to the multiplicative inverse of each point inertance (force excitation and acceleration measurement at the same point) as described above. The results are the apparent mass transfer functions $H_k^{AMF}(f)$ (equation 2.8). Generally, the apparent mass transfer functions are not equal to the diagonal elements $M_{k,k}(f)$ of the apparent mass matrix $\mathbf{M}(f)$. The former are the multiplicative inverse of the inertances $I_{k,k}(f)$ (equation 2.7) considering no coupling. The latter are the result of a matrix inversion of a dense inertance matrix. Both are equal, if a diagonal inertance matrix is used.

Force calculation

For each element $M_{k,l}(f)$ of the apparent mass matrix, a transfer function is built by combining the corresponding values of all frequencies (Figure 2.10). Impulse responses $h^{\mathrm{M}}_{k,l}(t)$ are calculated from the resulting transfer functions by means of the inverse Fourier transform. Afterwards, the acceleration signals are filtered with the related impulse responses and summed. The result is the force signal at position k:

$$\tilde{F}^{IFD}_k(t) = \sum_{l=1}^{L} a_l(t) * h^{\mathrm{M}}_{k,l}(t). \tag{2.17}$$

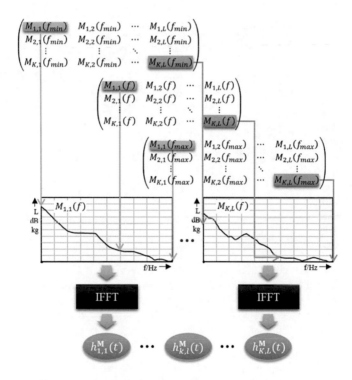

Figure 2.10: *For each element of the apparent mass matrix, a transfer function is built by combining the corresponding values of all frequencies. Impulse responses are calculated from the resulting transfer functions by means of the inverse Fourier transform.*

Structure-borne sound synthesis

The structure-borne sound can be calculated by filtering the indirectly determined forces with the impulse responses related to the ATFs. The total structure-borne sound synthesis is equal to the sum of all paths:

$$\tilde{p}_i(t) = \tilde{F}_k^{IFD}(t) * h_{k,i}^{ATF}(t). \tag{2.18}$$

The flow chart of the transfer path synthesis (TPS) is shown in Figure 2.11.

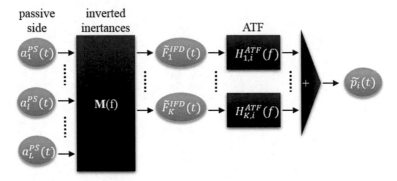

Figure 2.11: *Structure-borne sound synthesis using indirectly-determined forces.*

2.4.3 Regularization

Mathematical techniques like regularization, e.g., in the form of a singular value rejection are widely used in the case where the condition number of the inertance matrix is high [64, 65, 66]. The regularization can improve the result by preventing an excessive force overestimation, but it does not guarantee correct results [65, 66].

If the condition number of the inertance matrix at a frequency is high, then small measurement noise in the operational accelerations can lead to big errors in the indirectly determined operational forces at this frequency. This error amplification is the reason for overestimated forces.

The singular value rejection method uses the fact that a matrix inversion can be performed by the singular value decomposition. The singular value decomposition is a factorization of a matrix $\mathbf{X}$ into three matrices where $\mathbf{U}$ and $\mathbf{V}$ are unitary matrices containing the singular vectors and the diagonal matrix $\mathbf{S}$ contains non-negative real numbers called singular values.

$$\mathbf{X} = \mathbf{U} \cdot \mathbf{S} \cdot \mathbf{V}^T \qquad (2.19)$$

The singular value decomposition is used for computing the inverse of a matrix.

$$\mathbf{X}^{-1} = \mathbf{V} \cdot \mathbf{S}^{-1} \cdot \mathbf{U}^T \qquad (2.20)$$

The condition number of a matrix is high, if there is at least one singular value which is very small compared to the largest singular value. The inversion of a very small singular value gives a high number which amplifies an error in the measured acceleration vector (equation 2.11). In many cases it is useful to reject these small singular values during the matrix inversion. That means, new matrices are used for the inversion.

$$\tilde{\mathbf{X}}^{-1} = \tilde{\mathbf{V}} \cdot \tilde{\mathbf{S}}^{-1} \cdot \tilde{\mathbf{U}}^T \qquad (2.21)$$

They are a subset of the singular value decomposition matrices because small singular values and their related singular vectors are removed. The number of rejected singular values regulates the degree of regularization.

2.4.4 In-situ inverse force identification

An in-situ inverse force identification avoids the time-consuming disassembling of the source structure for the impact testing [67, 68, 69]. That means the inertance or mobility FRFs of the coupled system (Figure 2.12) are used in equation 2.11.

Figure 2.12: *For the in-situ inverse force identification the inertance FRFs of the coupled system are used. The engine is not removed.*

The resulting forces are equal to the blocked forces [67, 68], which, in general, are not equal to the operational forces. The explanation is that without disassembling an unknown portion of the force induced by the impact hammer acts on the source structure. This can lead to different inertance FRFs compared to the case with separated structures, and thereby to different determined forces. If the inertances of the source structure are sufficiently high compared to the inertances of the receiver structure, the difference between the two forces is negligibly small [68]. Then the impact hammer's force acts almost exclusively on the receiver structure and not on the source structure. This is often valid for elastically mounted sources. The forces based on the inertances of the coupled system must be combined with the (vibro-) acoustic transfer functions, which are also measured at the coupled system, in order to synthesize the total structure-borne sound share. The result is the same as when using operational forces and acoustic transfer functions of the uncoupled system [67]. This method is used for both engine [57] and tire-road noise [70, 71].

2.5 Two-port model

The indirect force determination can be performed in the case of a rigid connection between source and receiver structure as well as in the case of a resilient mounting. But mounts are not part of this transfer path method. In that case, it is not possible to gain any information about a mount although it is an important part of the transfer path with potential room for improvements [72].

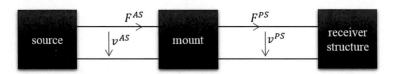

Figure 2.13: *Two-port model of a mount with source and receiver structure*

The two-port network theory, which is also called the quadripole theory, can be used to describe the transfer characteristics of a mount [73]. This theory, initially developed for the analysis of electrical circuits, can also be used for vibro-acoustical problems [74]. The source impedance is connected on the left side of the mount two-port and the structure is connected as a load impedance on the right side (Figure 2.13). The advantage of this model is that feedback between source, mount and structure is included.

Each mount is modeled as three independent single degree of freedom (SDOF) two-ports assuming that the directions of space are independent from each other [75]. Another assumption is that there is no cross-coupling between the different force excitation points. The idea behind two-port network theory is that a two-port can be completely described by four frequency depended two-port parameters [28].

$$\begin{pmatrix} F^{AS}(f) \\ F^{PS}(f) \end{pmatrix} = \begin{pmatrix} Z_{11}(f) & Z_{12}(f) \\ Z_{21}(f) & Z_{22}(f) \end{pmatrix} \cdot \begin{pmatrix} v^{AS}(f) \\ v^{PS}(f) \end{pmatrix} \tag{2.22}$$

If these parameters are known, the force induced into the passive structure can be calculated from the active-side velocity (or acceleration) and the load impedance $Z_L(f)$ [76].

$$F^{PS}(f) = \frac{Z_{21}(f)}{1 + \frac{Z_{22}(f)}{Z_L(f)}} \cdot v^{AS}(f) \tag{2.23}$$

The active-side velocity can be calculated from the acceleration signal which is measured with a triaxial accelerometer placed on the source near the mount. The load impedance is the point incrtance i.e. it is determined by impact measurements. Because feedback is modeled, predictions can be made about what happens if, for example, the load impedance is changed. In principle, the two-port parameters can be determined using test-rig data [75, 77] or FEA (finite element analysis) results [76, 78]. But in practice this is very time-consuming and elaborate [79, 80]. For example, adaptors must be crafted to connect the mount to the test rig. Often, the question arises whether this test rig data is transferable to the real situation of the mount or bushing in the car [81]. The two-port parameters can be determined in-situ [82]. But then two different measurements with varied receiver structure impedances are necessary. The structure impedance could be modified with an additional mass. In practice, creating a sufficient impedance variation is a challenge.

2.6 Mount stiffness method

The mount stiffness method [12] can be expressed as a simplification of the two-port method described above. Then the mount two-port consists of only one (complex-valued) mechanical impedance which contains the mount stiffness characteristic (Figure 2.14). If active-side and passive-side velocities are known the passive side force can be calculated.

$$F^{PS}(f) = Z^{MNT}(f) \cdot \left(v^{AS}(f) - v^{PS}(f) \right) \tag{2.24}$$

In this case the mechanical impedance of the structure is not necessary. In practice determining the active-side and passive-side velocities from measured acceleration signals is not the major challenge. Apart from a possible violation of the made simplifications one deals with a similar problem of determining reliable parameters in the form of mount stiffness profiles. The dynamic stiffness can also be calculated using FEA [78, 83]. On the other hand, compared to inverse force identification the mount stiffness method needs less measurement effort, especially if mount stiffness is known.

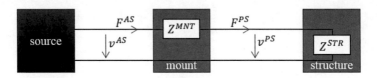

Figure 2.14: *The simple mount two-port exists only of one complex-valued impedance which corresponds to the mount stiffness.*

2.7 Mount attenuation function method

The mount attenuation function method (MAF), published as binaural transfer path analysis and synthesis (BTPA/BTPS) in 1999 [84], combines mount characteristics with the indirect impedance method [84, 85, 86, 20]. The mount attenuation function method is a compromise between two-port method and mount stiffness method. It is based on the same simplified two-port model containing only one complex impedance (see Figure 2.14). The difference is that no mount stiffness from test rig data is necessary. Instead, a mount attenuation is extracted from operational data. It is the frequency dependent relation between active and passive-side acceleration. In [87] it is shown that the mount attenuation function method is equal to the two-port method in the case of the simplified two-port model.

In practice the MAF method works well on weakly coupled system with soft mounts because of the underlying assumptions and simplifications. The two-port method and the simplifications mount attenuation and mount stiffness methods have in common that the cross-coupling of the structure is not considered although the indirect force determination shows better results in these cases if a dense inertance matrix is used.

2.7.1 Derivation of mount attenuation function method

[4] For each position and direction of space, the mount attenuation function is defined as transfer function between active-side (AS) and passive-side (PS) accelerations

---

[4]This part is a revised section taken from [59].

measured during operational conditions.

$$H_k^{MAF}(f) = a_k^{PS}(f) \oslash a_k^{AS}(f) \tag{2.25}$$

The assumption is made that there is no crosstalk within the mount. An impact hammer is not used to determine the attenuation for two reasons. First, it is usually not possible to induce sufficient energy for a good signal-to-noise ratio in the passive-side acceleration signal. Second, the attenuation depends on the load condition due to non-linear mount characteristics. For each load condition, like wide-open throttle or partial load, the transfer function is calculated from operational data. It is assumed that the mount characteristics are linear within each considered load condition. In principle, this is an operational transfer path analysis (OTPA) with only one input and one output signal (Figure 2.15).

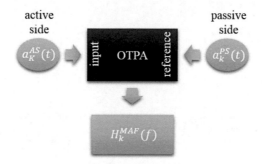

Figure 2.15: *The mount attenuation function (MAF) is the transfer function between active-side acceleration and its related passive-side acceleration. It is calculated from operational data.*

In the auralization process the mount attenuation suitable for the given load is selected and the active-side acceleration is filtered with the related impulse response:

$$\tilde{a}_k^{PS}(t) = a_k^{AS}(t) * h_k^{MAF}(t). \tag{2.26}$$

In the next step the resulting force is calculated without a matrix inversion and only considering the local body impedance as described in equation 2.8:

$$\tilde{F}_k^{MAF}(t) = a_k^{AS}(t) * h_k^{MAF}(t) * h_k^{AMF}(t). \tag{2.27}$$

The structure-borne noise share of a path can be calculated using the impulse response of the vibro-acoustic transfer function.

$$\tilde{p}_k^{MAF}(t) = \tilde{F}_k^{MAF}(t) * h_k^{ATF}(t) \tag{2.28}$$

The transfer path model is shown in Figure 2.16. Each transfer path consists of a sequence of three transfer functions: mount attenuation, apparent mass and vibro-acoustic transfer function. Although this approach is based on active-side accelerations with the aforementioned advantages, the coupling of the structure is ignored, which often leads to an overestimation of the forces. The influence of an uncorrelated source disturbing the passive-side acceleration signals can be eliminated by calculating the transfer function using correlation techniques. The passive-side acceleration at position k also contains correlated shares which are caused by the same source inducing forces at the other positions $(1 \cdots K)$. This crosstalk leads to an incorrect estimate of the mount attenuation.

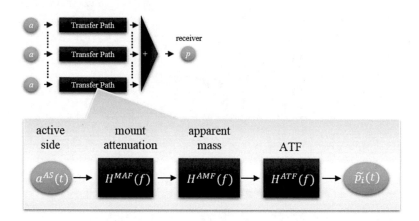

Figure 2.16: *A transfer path according to the mount attenuation function method consists of mount attenuation, apparent mass and acoustic transfer function.*

2.8 Transmissibility method - Operational transfer path analysis

The operational transfer path analysis (OTPA) is based only on operational data. Transfer functions of a multiple-input-multiple-output (MIMO) model as shown in Figure 2.17 are not derived from additional time-consuming measurements like in a conventional TPA. They are determined using correlation techniques from input signals $x_j(t)$ and receiver signals $y_i(t)$ which are simultaneously measured during operation. In principle, any kind of signal like acceleration, force or sound pressure can be used, if it characterizes the excitation of the sound source. If both source and receiver signals are responses of the physical mechanism of origin, the FRFs are referred to as transmissibilities. In this case it is a response-response model.

2.8.1 Multiple-input-multiple-output model

Operational transfer path analysis has become popular in the last decade [88], although correlation techniques of multiple-input-multiple-output systems have been published long before that [23]. But now powerful computers and digital multi-channel measurement systems are available, making OTPA a fast and easy method. On the other hand, OTPA has some limitations because the transfer functions are just a result of a mathematical optimization, which is not bound to a physical model or the principle of cause and effect. Missing paths, a high cross-coupling between measured input signals or highly coherent input signals, can therefore lead to a wrong path identification [89, 90, 91, 92]. Nevertheless, OTPA can be a helpful tool for system identification or sound quality troubleshooting if the limitations are taken into account.

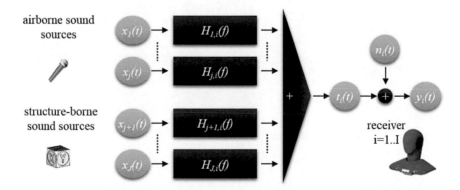

Figure 2.17: *Multiple-input-multiple-output (MIMO) model for operational transfer path analysis (OTPA)*

2.8.2 Derivation of OTPA

[5] In the OTPA algorithm the recorded time signals $x_j(t)$ and $y_i(t)$ are divided into overlapping blocks, multiplied with an analysis window (typically a Hann window) and then transformed to $X_j(f,m)$, $j = 1..J$, and $Y_i(f,m)$, $i = 1..I$, using the fast Fourier transform (FFT) with frequency index f and block index $m = 1..M$. The matrix notation allows a short and elegant description. For every frequency f the input quantities are combined to the matrix $\mathbf{X}(f)$ containing the values for all input sensors and all blocks.

$$\mathbf{X}(f) = \begin{pmatrix} X_1(f,1) & \cdots & X_1(f,M) \\ \vdots & \ddots & \vdots \\ X_J(f,1) & \cdots & X_J(f,M) \end{pmatrix} \tag{2.29}$$

For each receiver signal the values are arranged as a row vector.

$$\mathbf{Y}_i(f) = \begin{pmatrix} Y_i(f,1) & \cdots & Y_i(f,M) \end{pmatrix} \tag{2.30}$$

The transfer functions $H_{j,i}(f)$ from input j to output i are also arranged as a row vector.

$$\mathbf{H}_i(f) = \begin{pmatrix} H_{1,i}(f,1) & \cdots & X_{J,i}(f,M) \end{pmatrix} \tag{2.31}$$

The measured receiver signal $y_i(t)$ consists of two parts: $t_i(t)$ is correlated and $n_i(t)$ is uncorrelated to the input signals. The first part is transmitted from the considered sources via their transfer paths to the receiver and the second part has its origin in other uncorrelated sources or disturbing noise.

$$\mathbf{Y}_i(f) = \mathbf{T}_i(f) + \mathbf{N}_i(f) \tag{2.32}$$

The part which is correlated to the input signals can be synthesized by the sum of the filtered input signals. Filtering with the impulse response is equivalent to a multiplication with the transfer function in the frequency domain. This can be expressed for every frequency by the following system of equations.

$$\mathbf{Y}_i(f) = \mathbf{H}_i(f) \cdot \mathbf{X}(f) + \mathbf{N}_i(f) \tag{2.33}$$

The number of blocks M needed to solve this system of equations is at least equal to the number of input signals. In the case of an over-determined system the solution is found using the pseudo-inverse [62].

$$\mathbf{H}_i(f) = \mathbf{Y}_i(f) \cdot \mathbf{X}(f)^+ \tag{2.34}$$

Eq. 2.34 is the best solution in the least-square sense.

$$\min \|\mathbf{Y}_i(f) - \mathbf{H}_i(f) \cdot \mathbf{X}(f)\|_2 \tag{2.35}$$

---

[5]This section is a revised version taken from [93].

OTPA determinates the transfer functions in such a way that the deviation between receiver measurement and synthesis is as small as possible (in the least-square sense). Because $x_j(t)$ and $n_i(t)$ are uncorrelated there is no linear dependence between $\mathbf{X}$ and $\mathbf{N}_i$. With a sufficient number of blocks the influence of $n_i(t)$ is reduced due to averaging. The sum of the filtered input signals is the synthesis

$$\hat{t}_i^{OTPA}(t) = \sum_j h_{j,i}(t) * x_j(t), \tag{2.36}$$

where $h_{j,i}(t)$ is the corresponding impulse response to $H_{j,i}(f)$. The difference between the measured receiver signal and OTPA synthesis is the residual, which is the uncorrelated part in relation to the input signals.

$$\hat{n}_i^{OTPA}(t) = y_i(t) - \hat{t}_i^{OTPA}(t) \tag{2.37}$$

2.9 Model based approaches

In [94] a transfer path analysis method for an engine TPA is published which is based on parametric load models. The method is referred to as "operational path analysis with eXogenous inputs (OPAX)"' [94]. It is a scalable approach supporting different kinds of possible models. For structure-borne paths the range goes from a simple mount stiffness method to a complete indirect force measurement with matrix inversion. In between SDOF or multi-band models can be used for the dynamic stiffness of the mounts. The user must choose the model that best suits the actual problem.

Airborne transfer paths need a parametric model which calculates the volume velocity from sound pressure signals measured at the source during operation. OPAX is a hybrid approach combining operational data and measured FRFs. The goal is to avoid extensive FRF measurements which were necessary for a conventional TPA. At least, vibro-acoustic transfer functions from the loads to the receivers are necessary to calculate the sound shares. Additional FRFs from force excitation positions to indicator accelerations can be included to improve the results. These FRFs are nothing else than structure inertances, which would be used in an inverse force identification as described above. The FRFs of the uncoupled systems are recommended in [94]. The difference compared to the standard inverse force identification is the fact that depending on the chosen mount models the number of the necessary FRFs can be significantly reduced.

In the case of engine structure-borne sound, parameters of dynamic stiffness mount models are determined by solving an overdetermined linear equation system in the frequency domain. These equations contain operational data as well as measured

FRFs. The unknowns are the parameter of underlying models which are calculated in a least-squares sense. The concrete structure and composition of the equations depends on the used parametric models. Examples can be found in [94]. The calculations are based on complex-valued order data. That means magnitude and phase values of the engine orders are extracted from measured active and passive-side accelerations during operation. The loads are the forces induced by the engine into the car body during operation.

For example, a single degree of freedom mount stiffness model needs only three parameters: dynamic mass, damping and static stiffness. An overdetermined system of equations can be set up by evaluating many orders at different RPM values. This is opposite to the case of the OTPA algorithm where a linear equation system is solved independently for each frequency with one unknown per frequency and transfer function. An optimization for many frequencies together (like in OPAX) promises a more robust estimation of the mount characteristics, but of course the parametric model must fit the actual problem sufficiently.

A disadvantage of the OPAX algorithm is the complexity of choosing adequate parametric load models. A very simple model with a few parameters could be too simple. Too many parameters in conjunction with only a few FRFs and low variance in the input data could lead to ill-conditioned problems and therewith to wrong results. Furthermore, it is a pure frequency domain algorithm. The capability of synthesizing time signals which would allow for an appropriate auralization has not been mentioned in the literature.

2.10 Discussion

The operational transfer path analysis allows a fast and easy to use evaluation of the tire-road noise during coast-down measurements while the engine is switched off [95]. The tire-road noise, as well as the airborne and structure-borne sound shares, can be auralized without wind noise. The engine must be switched off or at least idling, otherwise unwanted engine noise shares would be present in the tire-road noise auralization. The methodology needs to be extended to also process measurements under dynamic driving conditions with a running engine but without engine sound shares. First, coast-down measurements without the engine running have no practical relevance to customers' driving experience. Second, the tire-road noise can vary due to the different load and drive torque under dynamic driving conditions. In this thesis an approach is presented which performs a cross-talk-cancellation to eliminate the influence of the engine on the tire-road noise synthesis.

A transfer path analysis method, which is widely used in the development process of a vehicle, is the inverse force identification. It is an experimental approach to identify dominant sources and transfer paths of a disturbing or unwanted noise pattern. It is used for chassis as well as powertrain issues. If there are several source structure attachment points in very close proximity, the inverse force identification reaches its limits. In these cases, ill conditioned inertance FRF matrices can lead to wrong results [13]. This method does not include mounts in the TPA model. In fact, mounts are an important part of the structure-borne transmission path which can give a lot of possibilities for improving the transmission behavior. Close to the end of the development process the only parts which can be modified with reasonable effort are the engine mounts. Then it is usually too late for local stiffness modification of the car body. Therefore, mounts should be modeled in a transfer path analysis and synthesis.

The two-port method models the relation between source, mount and structure, but determining the necessary two-port parameters of a mount on a special test rig is very time-consuming and elaborate. Setting the correct preload is difficult. The dynamic behavior of the mount adaptors can also lead to wrong results. So it is questionable whether the parameters determined on test rigs are transferable to the real situation of the mount in the car. The mount stiffness method, which is a simplification of the two-port method, deals with the same challenges. The mount attenuation function method is, in principle, based on the same model, but it is not dependent on test rig data. The mount attenuation is calculated in-situ from operational data. The disadvantage of these methods is that a coupling of the different directions and positions of the structure is neglected, although the inverse force identification method can consider it using matrix inversion. The OPAX method can estimate mount stiffness data based on a parametric model but the coupling of the receiver structure is not considered. The focus of OPAX lies more on reducing measurement effort than on increasing the quality of the estimated forces. It is a frequency domain method, so an auralization is not available.

So far, a method is missing which combines the advantages of the mount attenuation method and the inverse force identification method. These are determining mount properties from in-situ measurements and considering the strong coupling of the receiver structure while an auralization of the results is possible. Another aspect is parametric mount models which give a better physical insight into the system and allow for a parameter study. This helps with finding the best configuration or studying the influence of manufacturing tolerances on the structure-borne sound.

Chapter 3

Engine TPA - Parameterizing mount models based on in-situ measurements

In a modern vehicle a high standard of airborne insulation is the reason that the structure-borne sound share caused by the engine is an important part of the driving noise. Tools and methods for analyzing structure-borne sound transmission are very helpful in the development process.

In this chapter a new approach is presented which unites the benefits of the inverse force identification with matrix inversion and the advantages of the mount attenuation function method. It is called the effective mount transfer function method. Mount characteristics are calculated in-situ from operational data considering the crosstalk on the structure. The individual steps to determine the effective mount transfer functions are explained in the first section. These transfer functions are an ideal starting point for estimating parameters of mount models in the following section. Physically motivated parameter models with only a few parameters are used to avoid overfitting and therewith meaningless values.

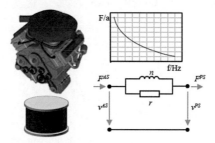

Figure 3.1: *Parametric mount models can be used in engine TPA.*

Auralization is a key feature for the assessment of perceived sound quality, but auralization artifacts can distract from the proper sound pattern and should be avoided. Methods and strategies are discussed to detect and reduce these artifacts. Finally, a new approach is shown for how input signals measured on a test rig can be properly used in a transfer path synthesis model of a vehicle.

3.1 Effective mount transfer function method

[1] For the derivation of the effective mount transfer function method it is assumed that the source is nonreactive and impresses a velocity into the mount. Furthermore, the coupling within the source structure and within the mount is negligible, but the coupling of the receiver structure must be considered. Additionally, the inertance of the mount is clearly higher than the inertance of the receiver structure because in general the mount material (rubber) is much softer than the structure material (steel or aluminum). The geometry of the car body at the mount interface is usually designed as a stiff structure and not as a very thin and elastic plate. That means a force induced into the mount at a given position can be determined using only the active-side acceleration at the same position and in the same direction. On the contrary, the force induced into the receiver structure cannot be calculated using only the related passive-side acceleration due to the high coupling. But this force can be calculated considering only the related active-side acceleration.

This can be expressed by a combination of two-ports representing the mounts and by an n-pole for the receiver structure (Figure 3.2). A mount is usually modeled by three two-ports, one for each spatial direction [76]. Because of the assumptions made above, the mount two-port has a simple configuration consisting of only one complex mechanical impedance $Z_k(f)$. The impedance of each mount two-port with the index k can be calculated from active- and passive-side velocities and passive-side force.

$$Z_k(f) = \frac{F_k^{PS}(f)}{v_k^{AS}(f) - v_k^{PS}(f)} \tag{3.1}$$

In practice, accelerations are measured instead of velocities so that, in what follows, the inertance

$$I_k^{MNT}(f) = \frac{a_k^{AS}(f) - a_k^{PS}(f)}{F_k^{PS}(f)} \tag{3.2}$$

of each mount two-port will be used. All mount inertances can be combined to a

---

[1]This part is a revised section taken from [59].

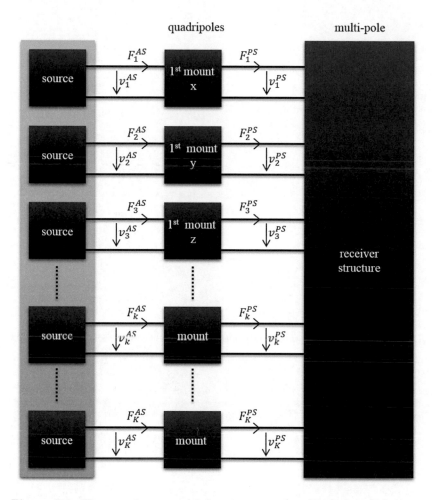

Figure 3.2: *The mounts are modeled as two-ports (quadripoles) and the receiver structure as multi-pole considering its strong coupling.*

diagonal matrix because of the assumptions made above:

$$\mathbf{I}^{MNT}(f) = \begin{pmatrix} I_1^{MNT}(f) & 0 & \cdots & 0 \\ 0 & I_2^{MNT}(f) & \cdots & 0 \\ \vdots & \vdots & \ddots & \vdots \\ 0 & 0 & \cdots & I_K^{MNT}(f) \end{pmatrix}$$

$$= \mathrm{diag}(I_1^{MNT}(f), \cdots, I_K^{MNT}(f)) \tag{3.3}$$

Equation 3.2 can be transformed into a matrix-vector notation with the vector of passive-side forces $F_k^{PS}(f)$ and the vector of the active and passive-side accelerations $a_k^{AS}(f)$ and $a_k^{PS}(f)$:

$$\mathbf{F}^{PS}(f) = \mathbf{I}^{MNT}(f)^{-1} \cdot (\mathbf{a}^{AS}(f) - \mathbf{a}^{PS}(f)) \tag{3.4}$$

Equation 2.12 is used to replace the passive-side accelerations with the inertance matrix of the structure $\mathbf{I}^{STR}(f)$ and the force vector:

$$\mathbf{F}^{PS}(f) = \mathbf{I}^{MNT}(f)^{-1} \cdot (\mathbf{a}^{AS}(f) - \mathbf{I}^{STR}(f) \cdot \mathbf{F}^{PS}(f)) \tag{3.5}$$

This equation can be transformed into a new equation describing the relation between active-side accelerations and passive-side forces:

$$\mathbf{F}^{PS}(f) = \left(\mathbf{I}^{MNT}(f) + \mathbf{I}^{STR}(f)\right)^{-1} \cdot \mathbf{a}^{AS}(f) \tag{3.6}$$

The inertance matrices of mount and structure must be summed before the matrix inversion:

$$\mathbf{I}^{MNT+STR}(f) = \mathbf{I}^{MNT}(f) + \mathbf{I}^{STR}(f) \tag{3.7}$$

Because the inertances of the mounts (diagonal matrix) are higher than the inertances of the source structure this matrix can be approximated by a diagonal matrix. Hence, the inverse of $\mathbf{I}^{MNT+STR}(f)$ can be expressed by a diagonal matrix $\mathbf{D}(f)$:

$$\mathbf{D}(f) \cong \mathbf{I}^{MNT+STR}(f)^{-1} \tag{3.8}$$

Equation 3.6 can be rewritten to a simpler relation between active-side acceleration and passive-side force:

$$\mathbf{F}^{PS}(f) \cong \mathbf{D}(f) \cdot \mathbf{a}^{AS}(f) \tag{3.9}$$

The diagonal matrix $\mathbf{D}(f)$ contains the effective mount transfer function $H_k^{MTF}(f)$ which represents the relationship between active-side acceleration and induced force into the structure through the mount:

$$H_k^{MTF}(f) = F_k^{PS}(f) \oslash a_k^{AS}(f)$$

$$= \frac{1}{I_k^{MNT}(f) + I_{k,k}^{STR}(f)} \tag{3.10}$$

It is called "'effective'" because it describes the mount in the actual assembly situation containing the properties of both mount and structure. Using the velocity instead of the acceleration, this is equal to a transfer impedance $Z_k^{MTF}(f)$ of the coupled system:

$$Z_k^{MTF}(f) = H_k^{MTF}(f) \cdot j\omega = F_k^{PS}(f) \oslash v_k^{AS}(f) \tag{3.11}$$

In general, the transfer impedance of the coupled system is not equal to the transfer impedance of the two-port with short-circuited output poles with $v_k^{PS}(f) = 0$ which is equal to $Z_k(f)$. But, if the passive-side velocity is very small compared to the active-side velocity, the transfer impedance of the coupled system is approximately equal to the transfer impedance of the two-port. In this case, the induced force depends only on the active-side velocity and the mount impedance.

For the mount attenuation function (see section 2.7), the correlation between active-side and passive-side acceleration is evaluated ignoring the coupling of the structure. In the new approach, the correlation between the active-side acceleration and the force acting on the passive-side is calculated. This force is determined indirectly using all passive-side accelerations now considering the coupling of the structure.

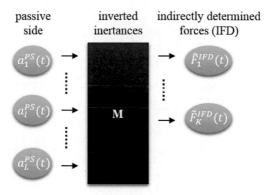

Figure 3.3: *The first step of calculating effective mount transfer functions is synthesizing indirectly determined forces.*

The approach consists of three steps. First, an indirect force determination is performed based on acceleration sensors applied on the passive-side and inertance measurements (Figure 3.3 and see section 2.4). In the case of a vehicle with massive engine and soft mounts the assumption is fulfilled that there is no need to disassemble the source engine for the inertance measurements. The force of the impact hammer is acting almost entirely on the receiver structure and not on the engine. Furthermore, a separated engine could lead to different static loads of the structure and therefore to an unwanted change of the system. By inverting a dense inertance matrix, the crosstalk of the accelerations measured on the passive-side, which is caused by the strong coupling, is compensated. As already mentioned, these accelerations can still contain other noise components like tire-road noise.

In the second step, effective mount transfer functions are determined from operational data. For a given load condition, the force signals are synthesized according to the Indirect Force Determination. In the operational measurements, passive-side accelerations and active-side accelerations are recorded simultaneously so that for each direction and mount position a transfer function is determined between the active-side acceleration and the synthesized force signal using correlation techniques (Figure 3.4). This corresponds to an OTPA (see section 2.8) with one input and one output signal. The output signal, though a result of a calculation, is based on operational data. These first two steps are performed for each load condition of interest to accommodate the non-linear behavior of mounts.

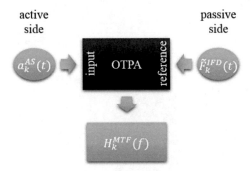

Figure 3.4: *The effective mount transfer function (MTF) at the position k is calculated from operational data using the active-side acceleration and the indirectly determined force.*

Finally, the active-side accelerations are filtered with the impulse responses of the effective mount transfer functions suitable for the load condition in order to calculate the operational forces:

$$\tilde{F}_k^{MTF}(t) = a_k^{AS}(t) * h_k^{MTF}(t) \tag{3.12}$$

These forces no longer contain noise components that are uncorrelated to engine vibrations. The structure-borne sound share can be calculated from the forces with the acoustic transfer functions as before (Figure 3.5).

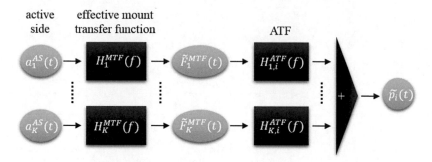

Figure 3.5: *Structure-borne synthesis based on effective mount transfer function (MTF) and (vibro-) acoustic transfer function (ATF) using active-side acceleration signals.*

If transfer paths without a mount between source and receiver structure should be modeled additionally, these paths can be included by simply adding their structure-borne sound contribution based on inverse force identification (Figure 2.11). In other words, the operational forces of paths containing a mount are calculated with effective mount transfer functions, and paths with a direct connection use the indirectly-determined forces of step one.

The assumptions for the derivation of the effective mount transfer function method are summarized in table 3.1. The effective mount transfer function describes the mount in its actual assembly situation. In many TPA cases, this is exactly what is wanted. There is no need to deal with typical challenges of measuring mount parameters on a test rig, like setting correct preloads or constructing mount adapters.

Table 3.1: *Assumptions*

	Assumption / Approximation	Comment
1	source is nonreactive and impresses a velocity into the mount	valid for engines, proofed in previous experiments
2	coupling within the source structure and within the mount is negligible	necessary assumption of the model, approximately fulfilled for typical engine mounts
3	coupling of the receiver structure must be considered	strong coupling of a car body can be validated by impact measurements
4	the inertance of the mount is clearly higher than the inertance of the receiver structure	in general, the mount material (rubber) is much softer than the structure material (steel or aluminum). The geometry of the car body at the mount interface is usually designed as a stiff structure.
5	a force induced into the mount at a given position can be determined using only the active-side acceleration at the same position and in the same direction	conclusion from assumption 1-4
6	a force induced into the receiver structure cannot be calculated using only the related passive-side acceleration due to the high coupling	conclusion from assumption 3
	active-side accelerations are less disturbed by other sources	in the case of high engine mass and soft mounts (typical situation of a combustion engine) other sound sources do not affect the active-side engine accelerations

3.2 Physically motivated parameter models

3.2.1 Kelvin-Voigt model

[2] With the effective mount transfer function no predictions can be made for the case that the source is mounted on another receiver structure with different inertances. Therefore, the parameters of the two-port are necessary. In the considered case of a simple two-port, only the impedance $Z_k^{MNT,MTF}(f)$ must be estimated (see Figure 3.6).

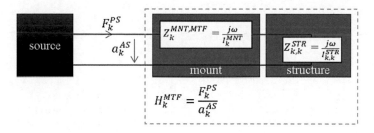

Figure 3.6: *The impedance of the two-port is calculated from the effective mount transfer function and the structure inertance.*

The impedance of the mount two-port $Z_k^{MNT,MTF}(f)$ is calculated from the effective mount transfer function $H_k^{MTF}(f)$ and the structure inertance $I_{k,k}^{STR}$.

$$Z_k^{MNT,MTF}(f) = \frac{j\omega}{\frac{1}{H_k^{MTF}(f)} - I_{k,k}^{STR}(f)} \tag{3.13}$$

This equation is derived from equation 3.10.

$$H_k^{MTF}(f) = \frac{1}{I_k^{MNT}(f) + I_{k,k}^{STR}(f)}$$
$$\frac{1}{H_k^{MTF}(f)} = I_k^{MNT}(f) + I_{k,k}^{STR}(f) \tag{3.14}$$

$$I_k^{MNT}(f) = \frac{1}{H_k^{MTF}(f)} - I_{k,k}^{STR}(f) \tag{3.15}$$

---

[2]This part is a revised and extended section taken from [87].

The tow-port impedance $Z_k^{MNT,MTF}(f)$ can be calculated from the mount inertance (see also equation 3.2).

$$Z_k^{MNT,MTF}(f) = \frac{j\omega}{I_k^{MNT}(f)} \qquad (3.16)$$

Finally, equation 3.13 is a combination of equation 3.15 and equation 3.16. The next step is using a parametric description for the two-port impedance $Z_k^{MNT,MTF}(f)$. For a simple rubber mount, the stiffness and viscous damping can be estimated by parameterizing a Kelvin-Voigt-model (Figure 3.7). A Kelvin-Voigt material is a viscoelastic material, which is named after Lord Kelvin and after Woldemar Voigt [96]. It has the properties of viscosity and elasticity. The parameters of viscous damping factor r with unit $N/(m/s)$ and resilience n with unit m/N are chosen so that the impedance of the model $Z_k^{KV}(f)$ fits best to $Z_k^{MNT}(f)$:

$$Z_k^{KV}(f) = r + \frac{1}{j\omega n} \qquad (3.17)$$

The corresponding flow chart diagram is shown in Figure 3.8. Starting with initial model parameters, a parametric two-port impedance is calculated. The mount parameters are determined so that the impedance of the model fits best to the mount impedance calculated from the effective mount transfer function.

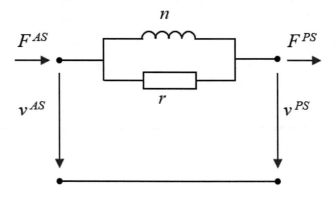

Figure 3.7: *Kelvin-Voigt model, named after Lord Kelvin and after Woldemar Voigt [96], consists of a spring (stiffness) and a resistor (viscous damping).*

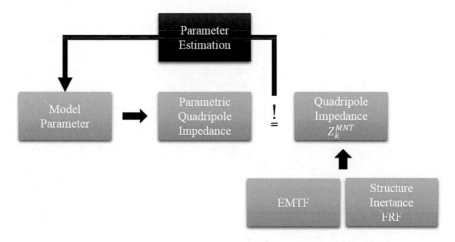

Figure 3.8: *This flow chart diagram shows how to estimate mount parameters from the effective mount transfer function and the structure inertance.*

3.2.2 Kelvin-Voigt model with mass

[3] For engine mounts the simple Kelvin-Voigt-model can be used up to 200 Hz. The mount mass has to be taken into account to extend the usable frequency range up to 1 kHz. Then the mount two-port does not consist of only one impedance but is more complex (Figure 3.9).[4]

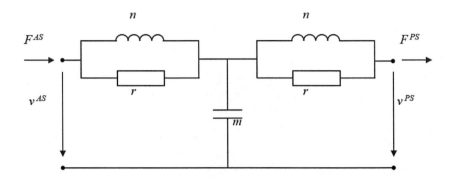

Figure 3.9: *Kelvin-Voigt model with mount mass*

The effective mount transfer function can be approximated by a Kelvin-Voigt-model with mass m

$$H_k^{KVM}(f) = \left(\frac{I_k^{KV} \cdot \left(I_k^{KV} + I_k^{STR} \right)}{I_k^{Mass}} + 2 \cdot I_k^{KV} + I_k^{STR} \right)^{-1} \tag{3.18}$$

with

$$I_k^{Mass} = \frac{1}{m} \tag{3.19}$$

and

$$I_k^{KV} = \frac{j\omega}{r + \frac{1}{j\omega n}} \tag{3.20}$$

The parameters m, n and r are chosen so that the result fits best to the calculated effective mount transfer function $H_k^{MTF}(f)$.

---

[3]This part is a revised section taken from [87].
[4]In this case it is not possible to calculate the mount impedance using equation 3.13 because of the mass in the shunt circuit.

3.3 Physically motivated models considering directional coupling within a mount

3.3.1 Considering directional coupling

[5] So far, the assumption was made that the spatial directions of a mount are independent of each other. Each direction was modeled by one Kelvin-Voigt model. Now the mount model will be extended so that an active-side force in one direction results in passive-side forces in all three dimensions. This directional coupling can be considered by generalizing the two-port theory to n-port theory. For this case, the effective mount transfer matrix, which expresses passive-side forces by active-side accelerations, is wanted.

The relation between the passive-side force vector and the active-side as well as the passive-side velocity vectors is characterized by the mount transfer impedance matrix and the mount output impedance matrix.

$$\mathbf{F}^{PS}(f) = \mathbf{Z}^{PS,AS}(f) \cdot \mathbf{v}^{AS}(f) + \mathbf{Z}^{PS,PS}(f) \cdot \mathbf{v}^{PS}(f) \tag{3.21}$$

The same equation can be expressed by using accelerations, which can be easily measured in practice. For a shorter notation, the frequency parameter f is omitted.

$$\mathbf{F}^{PS} = \mathbf{Z}^{PS,AS} \cdot \frac{1}{j\omega} \cdot \mathbf{a}^{AS} + \mathbf{Z}^{PS,PS} \cdot \frac{1}{j\omega} \cdot \mathbf{a}^{PS} \tag{3.22}$$

The main equation of the indirect force determination gives the relation between passive- side forces and passive-side accelerations.

$$\mathbf{a}^{PS} = \mathbf{I}^{STR} \cdot \mathbf{F}^{PS} \tag{3.23}$$

This equation can be inserted in equation 3.22 to eliminate the passive-side acceleration vector.

$$\mathbf{F}^{PS} = \mathbf{Z}^{PS,AS} \cdot \frac{1}{j\omega} \cdot \mathbf{a}^{AS} + \mathbf{Z}^{PS,PS} \cdot \frac{1}{j\omega} \cdot \mathbf{I}^{STR} \cdot \mathbf{F}^{PS} \tag{3.24}$$

The next two steps solve the equation for $\mathbf{F}^{PS}$.

$$\mathbf{F}^{PS} - \mathbf{Z}^{PS,PS} \cdot \frac{1}{j\omega} \cdot \mathbf{I}^{STR} \cdot \mathbf{F}^{PS} = \mathbf{Z}^{PS,AS} \cdot \frac{1}{j\omega} \cdot \mathbf{a}^{AS} \tag{3.25}$$

$$\mathbf{F}^{PS} = \left(1 - \mathbf{Z}^{PS,PS} \cdot \frac{1}{j\omega} \cdot \mathbf{I}^{STR}\right)^{-1} \cdot \mathbf{Z}^{PS,AS} \cdot \frac{1}{j\omega} \cdot \mathbf{a}^{AS} \tag{3.26}$$

---

[5]This part has been previously published in [97].

From this follows the effective mount transfer matrix.

$$\mathbf{H}^{MTF,model} = \left(1 - \mathbf{Z}^{PS,PS} \cdot \frac{1}{j\omega} \cdot \mathbf{I}^{STR}\right)^{-1} \cdot \mathbf{Z}^{PS,AS} \cdot \frac{1}{j\omega} \tag{3.27}$$

If the structure inertance is very low, that means the structure is very stiff compared to the mount, the effective mount transfer matrix is almost equal to the transfer impedance matrix of the mount.

For an auralization (time domain TPA) the time signals of the forces are necessary. The required impulse responses are calculated from the effective mount transfer function matrix. For each element of the matrix the values of all frequencies are combined to a vector, which is transformed to the time domain by the inverse Fourier transform. Then the active-side accelerations measured under operating conditions are filtered with the impulse responses and summed to yield the operational force.

$$\tilde{F}_k^{MTF,model}(t) = \sum a_j^{AS}(t) * h_{k,j}^{MTF}(t) \tag{3.28}$$

The next step is equal to a conventional force-based transfer path synthesis. The sound pressure shares at a given receiver are synthesized by filtering each force with the impulse response according to the related vibro-acoustic transfer function [84]. The results of all transfer paths are summed to yield the total structure-borne noise share.

3.3.2 Kelvin-Voigt models

[6] Now one mount is modeled by three Kelvin-Voigt models per direction. A stiffness and a damping parameter describes the coupling between one direction and another. The complete mount transfer matrix of a system consisting of three mounts is shown in the next equation.

$$\mathbf{Z}^{PS,AS} \cdot \frac{1}{j\omega} = \begin{pmatrix} I_{x,x}^{KV,1} & I_{x,y}^{KV,1} & I_{x,z}^{KV,1} & 0 & 0 & 0 & 0 & 0 & 0 \\ I_{y,x}^{KV,1} & I_{y,y}^{KV,1} & I_{y,z}^{KV,1} & 0 & 0 & 0 & 0 & 0 & 0 \\ I_{z,x}^{KV,1} & I_{z,y}^{KV,1} & I_{z,z}^{KV,1} & 0 & 0 & 0 & 0 & 0 & 0 \\ 0 & 0 & 0 & I_{x,x}^{KV,2} & I_{x,y}^{KV,2} & I_{x,z}^{KV,2} & 0 & 0 & 0 \\ 0 & 0 & 0 & I_{y,x}^{KV,2} & I_{y,y}^{KV,2} & I_{y,z}^{KV,2} & 0 & 0 & 0 \\ 0 & 0 & 0 & I_{z,x}^{KV,2} & I_{z,y}^{KV,2} & I_{z,z}^{KV,2} & 0 & 0 & 0 \\ 0 & 0 & 0 & 0 & 0 & 0 & I_{x,x}^{KV,3} & I_{x,y}^{KV,3} & I_{x,z}^{KV,3} \\ 0 & 0 & 0 & 0 & 0 & 0 & I_{y,x}^{KV,3} & I_{y,y}^{KV,3} & I_{y,z}^{KV,3} \\ 0 & 0 & 0 & 0 & 0 & 0 & I_{z,x}^{KV,3} & I_{z,y}^{KV,3} & I_{z,z}^{KV,3} \end{pmatrix} \tag{3.29}$$

---

[6]This part has been previously published in [97].

All three mounts are included in one matrix, because then the complete coupling of the receiver structure can be considered by using a dense structure inertance matrix with the dimensions nine times nine in equation 3.29. Each mount is independent from the other mounts because of the zero entries in the transfer impedance matrix. For this mount model, the output impedance matrix is equal to the transfer impedance matrix with a negative sign.

$$\mathbf{Z}^{PS,PS} = -\mathbf{Z}^{PS,AS} \tag{3.30}$$

The mount parameters are determined so that the forces calculated by the model (equation 3.28) fits best to indirectly determined forces (Figure 3.10). Therefore, only in-situ measurements are needed to estimate the mount characteristics. The equivalent circuit diagram of one mount is shown in Figure 3.11. It consists of eighteen parameters, six per direction.

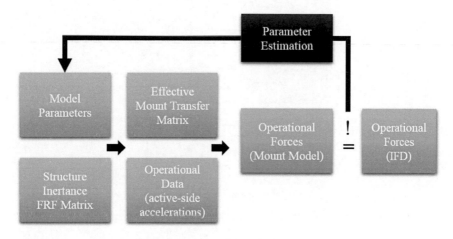

Figure 3.10: *This flow chart diagram shows how to estimate mount parameters from and the structure inertance FRF matrix.*

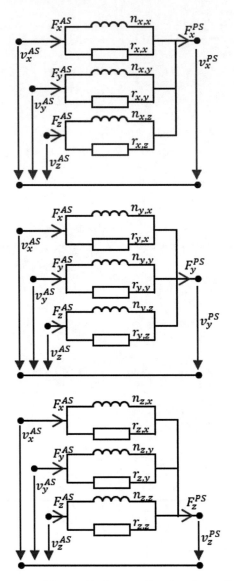

Figure 3.11: *Mount model considering a directional coupling based on three Kelvin-Voigt models per direction.*

3.4 Modeling of non-linear mounts

A mount can behave in a non-linear manner due to different temperatures, load or excitation amplitude [80]. Non-linearity can be either considered by a non-linear model or by combining piece-wise linearized models. In the latter case it is assumed that the mount is approximately linear if small changes occur.

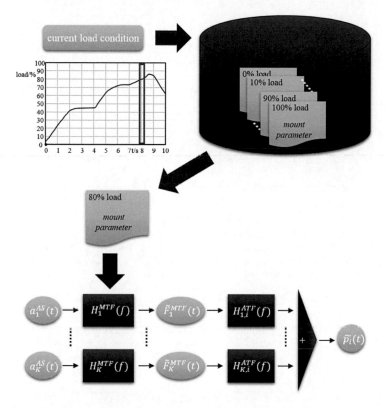

Figure 3.12: *A strategy is shown to consider a non-linear behavior of a mount. Mount parameter sets are determined independently for different load cases and stored in a database. In a block processing, according to a reference value like load, that parameter set is chosen for the current block, which is next to the actual load condition.*

A strategy for dealing with non-linear mounts is presented in Figure 3.12. Mount parameters are determined independently for different load cases, for example, from 10% to 100% with a step size of 10%. The parameter set of each load condition is stored in a database. There are two different strategies for a structure-borne sound synthesis. In general, a time block based approach is used. First, according to a reference value like load or accelerator pedal position, that parameter set is chosen for the current block, which is next to the actual load condition. The mount transfer functions are calculated according to this parameter set. The operational forces and the structure-borne sound synthesis are determined as described above. At the intersection of two different parameter sets the synthesis is cross-faded. In the second case, the mount parameters are linearly interpolated for each time block based on the adjacent parameter sets.

3.5 Avoiding auralization artifacts

The introduction chapter mentioned the importance of auralizing transfer path analysis results during the development process of a vehicle or a product (see section 1.3.7). In this context auralization artifacts would be counterproductive. Auralization artifacts are errors in the synthesis that are not present in the real noise. These are often small errors in respect to signal energy. If the focus lies only on diagrams containing order levels versus rpm, these artifacts can be neglected in most cases. However, they can be important if a person listens to the sound in a listening test. They can distract from the proper sound pattern and lead to a wrong or an inaccurate perception by the listener.

There are primary and secondary measures which can eliminate or reduce auralization artifacts. Primary measures use transfer path analysis approaches that produce inherently less auralization artifacts compared to other methods. Secondary measures are based on post-processing steps to detect and reduce artifacts.

3.5.1 Methods reducing auralization artifacts

Transfer path analysis and synthesis is based on a source-transmission-receiver model (see also Figure 1.5). This means, source signals or transfer functions can be the origin of auralization artifacts in the synthesis.

A transfer path synthesis, which is based on active-side accelerations, should be preferred, because active-side accelerations are less disturbed by other sound sources then passive-side accelerations. Using active-side accelerations leads to a clearer engine noise synthesis without unwanted noise shares from rolling tires or auxiliary components [59]. TPA approaches using active-side accelerations are the mount attenuation function method (see section 2.7) and the effective mount transfer function method (see section 3.1).

Given that operational input signals have been measured with care, the major remaining reason for auralization artifacts is errors like wrong peaks in the transfer functions of the transfer path model. A synthesis based on parametric mount models with a few parameters produces almost no artifacts because the resulting transfer functions are smooth and continuous. But the inverse force identification method based on matrix inversion calculates the transfer functions frequency by frequency, so that they can be discontinuous. Especially at frequencies with a high condition number of the inertance frequency response functions matrix, erroneous peaks can occur in the resulting transfer functions after matrix inversion. These peaks generate wrong tonal components. In many cases it is then helpful to use matrix inversion with regularization. A reasonable regularization reduces these artifacts. Another possibility could be using additional acceleration sensors to get overdetermined systems of equations with lower condition numbers.

Other artifacts can be created by errors in the measured inertance or vibro-acoustic frequency response functions. Indeed, they should be measured with utmost care. A critical point is achieving a sufficient high signal to noise ratio in all acceleration signals. This can be a challenge for sensors at positions which are quite far away from the impact position. Coherence analysis can help to identify critical positions and frequency regions. If during the impact testing several impacts are performed, the coherence between the impact hammer's force and the measured acceleration gives a hint about the quality of the inertance frequency response function. In some cases, it might be better to ignore the cross-coupling between positions with only a poor coherence. It makes more sense to set the corresponding values in the inertance matrix to zero than to use the measurement noise (see section 2.4). It is impossible to predict the consequences of errors in one inertance frequency response function in advance because the performed matrix inversion includes all inertances.

Remaining artifacts can be reduced in a post-processing step. Most artifacts are tonal components produced by wrong peaks in the transfer functions. The challenge is to distinguish between correct resonances and wrong peaks. In the case of the inverse force determination the unwanted peaks can be detected by evaluating the condition number of the inertance matrix. A peak in the condition number and narrow peaks in many transfer functions of the apparent mass matrix at the same frequency are good indicators for a local error. Interpolating the transfer function at this frequency can reduce the auralization artifacts. The algorithm is explained in Figure 3.13.

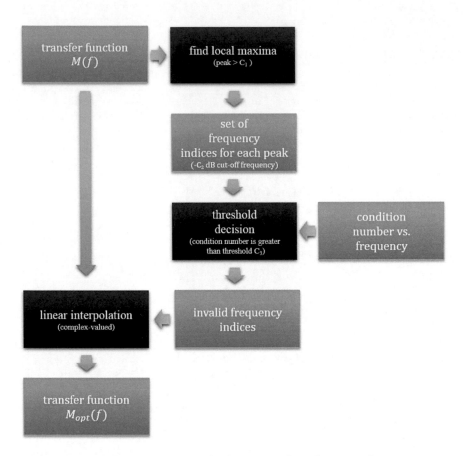

Figure 3.13: *Starting point: Transfer function and condition number.*
Step 1: Find local peaks in the transfer function.
Step 2: Threshold decision if condition number vs. frequency is above
a threshold at each peak position.
Step 3: Transfer function values at invalid frequency indices are re-
placed by interpolating between adjacent valid values.
Result: Optimized transfer function.

3.5.2 Example

The post-processing step to reduce auralization artifacts caused by wrong peaks is explained using an example. An indirect force determination with matrix inversion has been performed for a compact class car to calculate the structure-borne sound share of the tire-road noise. The operating condition is a coast-down measurement from 50 km/h to 30 km/h on a medium rough road. The acceleration sensors are placed at the four wheel hubs. The structure-borne sound synthesis is shown as a spectrogram in Figure 3.14. It contains auralization artifacts in the form of tonal noise components at higher frequencies, which are marked by small arrows in the figure.

The condition number of the inertance matrix is plotted versus frequency in Figure 3.15. At the frequencies where the auralization artifacts occur the values are considerably high, which indicates that the resulting determined forces and therewith the structure-borne sound shares are incorrect at these frequencies.

A second indicator is that there are narrow peaks in the apparent mass transfer functions - the result of the matrix inversion step - at the same frequencies. For example, one representative apparent mass transfer function is shown in Figure 3.16 as a green curve. The positions of the narrow peaks correspond to the high values of the matrix condition number. In a post processing step the auralization artifacts can be eliminated by calculating optimized transfer functions. At the identified frequencies with a high condition number and peak in the transfer function the values are replaced by interpolating between adjacent valid values. The narrow peaks can be removed with this method as it is shown in Figure 3.16 as a red curve.

The structure-borne synthesis based on the optimized apparent mass transfer functions no longer contains these auralization artifacts. The spectrogram is shown in Figure 3.17.

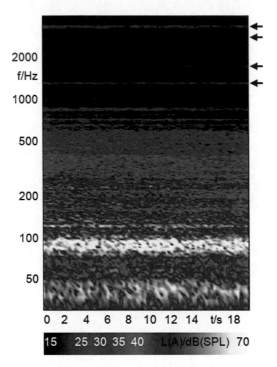

Figure 3.14: *In this example the structure-borne sound synthesis of the tire-road noise contains auralization artifacts in the form of tonal noise components at higher frequencies. The affected frequencies are marked by small arrows.*

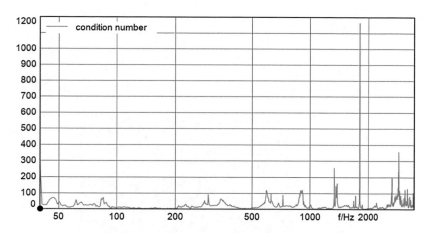

Figure 3.15: *The condition number of the inertance transfer function matrix vs frequency shows considerably high peaks at the frequencies where the auralization artifacts occur.*

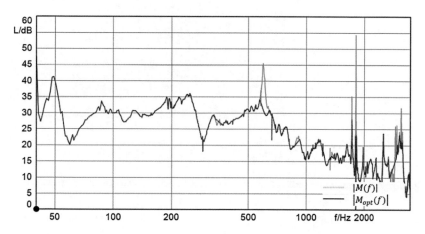

Figure 3.16: *As an example, one apparent mass transfer function $M(f)$ is selected to demonstrate suspicious narrow peaks at positions with a high condition number (green curve). An optimized transfer function $M_{opt}(f)$ can be calculated using interpolation (red curve).*

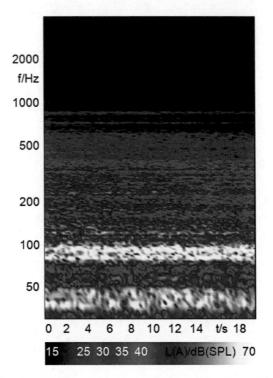

Figure 3.17: *The auralization artifacts can be removed using optimized transfer functions.*

3.6 Transfer path synthesis based on engine test rig data

3.6.1 Motivation

In the development process of an engine it is a great benefit if its interior noise can be predicted as early as possible. Judging the NVH (noise, vibration and harshness) comfort on the basis of a prototype engine without the need to install it in a real car can speed up the development process. Prototype modifications can be evaluated faster leading to a cost reduction. It is better to check early whether development targets are achieved, because late design changes are always expensive or limited. Meaningful predictions of sound quality cannot be achieved just by comparing order levels. It requires listening to an auralization of the engine noise at the driver's position [21].

Transfer path analysis and synthesis can be utilized to predict engine interior noise based on test rig data at an early stage of the development process. The engine is virtually installed into a car using transfer functions from an existing vehicle, e.g., the predecessor car, combined with signals measured on a test rig. The results are time signals, which can be analyzed and used in listening tests. On a test rig acceleration signals can be measured at the same position as in the car. Because of a different terminating impedance of the test rig and the vehicle these accelerations cannot be used directly in a TPA model derived from a real vehicle. First, the test rig accelerations must be adjusted to compensate the effect of the different receiver structure impedance (Figure 3.18). After that, the transfer functions of a vehicle can be used to predict the structure-borne sound of the engine.

3.6.2 Acceleration ratio based on reference measurements as correction function

One possibility is to calculate correction functions from two reference measurements. An identical engine is measured on the test rig as well as in the vehicle. The operating condition must be the same. For each acceleration signal a frequency dependent correction function is calculated from the ratio of the averaged spectra (according to equation 3.32) of the test rig data (A) and the vehicle data (B).

$$H_\Delta(f) = \frac{G_{a_{ref}^B}(f)}{G_{a_{ref}^A}(f)} \tag{3.31}$$

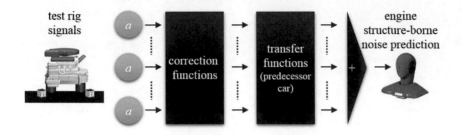

Figure 3.18: *Test rig accelerations of an engine must be adjusted to compensate the effect of the different receiver structure impedances of the test rig and vehicle. After that, the transfer functions of a vehicle can be used to predict the structure-borne sound of the engine.*

The averaged spectrum is calculated using the Fourier transform of M blocks of the time signal $s(t)$ using a Hann-window and 50% overlap.

$$G_s(f) = \sqrt{\frac{1}{M} \sum_{m=1}^{M} |S(f,m)|^2} \qquad (3.32)$$

Then, a new or modified engine needs to be measured on the test rig only. These acceleration signals are compensated by the correction functions of the previous step. Afterwards, they can be used for the transfer path synthesis based on the vehicle TPA model.

$$\hat{a}^B(f) = H_\Delta(f) \cdot a^A(f) \qquad (3.33)$$

Some assumptions must be made. The operating conditions like load are equal in all measurements. It is assumed that the source impedance of the engines is also equal. In addition, the engines do not have a different influence on the mounts. All directions and positions are independent of each other so that there is no coupling.

It is an advantage that this is an in-situ method where the mounts are in the same operating condition during the reference measurements to obtain the correction functions. But, on the other hand, one engine must be measured on a test rig as well as in a vehicle, which is time-consuming. One challenging task is to reproduce the same operating condition (temperature, load, etc.) as precisely as possible. The major drawback is that the correction function compensates only the magnitude and not the phase of the acceleration signal. In the transfer path synthesis a correct phase of each path signal is important because in the case of an engine highly correlated signals are superposed.

It can be shown that the correction function is independent from the operating condition. For each excitation position the equivalent circuit diagrams for test rig (A) and vehicle (B) consist of a vibration source, a source impedance Z_S and a load impedance Z_L (Figure 3.19). The velocity at the load impedance depends on the source velocity and the source and load impedance.

$$
\begin{aligned}
v_L^A(f) &= v_s(f) \cdot \frac{Z_S(f)}{Z_L^A(f) + Z_S(f)} \\
v_L^B(f) &= v_s(f) \cdot \frac{Z_S(f)}{Z_L^B(f) + Z_S(f)}
\end{aligned}
\tag{3.34}
$$

The velocity ratio of the cases test rig (A) and vehicle (B) depends only on the impedances and is independent from the actual source velocity.

$$
H_\Delta(f) = \frac{v_L^B(f)}{v_L^A(f)} = \frac{Z_L^A(f) + Z_S(f)}{Z_L^B(f) + Z_S(f)}
\tag{3.35}
$$

The correction function which is derived from the ratio of the averaged spectra is equal to the magnitude of the true correction function of the underlying model. The formula shows that it can be calculated independently from the operating condition (the source velocity) as long as the operating condition is equal in both cases. It is assumed that both systems behave linear.

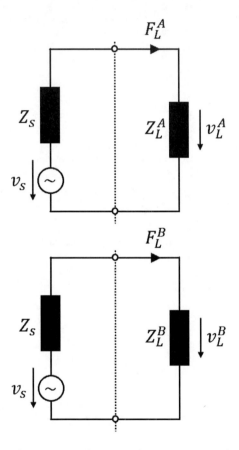

Figure 3.19: *Equivalent circuit diagrams for a source with two different load impedances A and B*

3.6.3 Impedance based method

A new method is presented which takes the inertance of the test rig and the car into account. Elaborate reference measurements of an engine are not necessary. Here, the structure inertances are measured with an impact hammer while the source is attached to the receiver. An equivalent circuit diagram is shown in Figure 3.20. In most cases this has already been performed for the vehicle during the transfer path analysis, if the in-situ inverse force identification method is applied (see section 2.4.4).

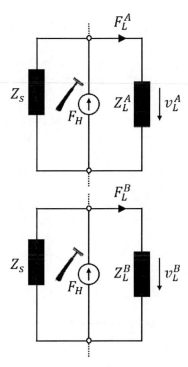

Figure 3.20: *Equivalent circuit diagrams for the impact measurement at the interface between a source and load impedance for cases A and B*

First, directions and positions are considered as independent of each other. No coupling is assumed. The mobility $Y(f) = v(f)/F(f)$ can be calculated from the inertance by multiplying with $j\omega$. In both cases the mobility at the interface between source and receiver structure depends on the source and load impedance.

The index SL indicates that this is the mobility of the assembled system composed of the source structure S and the receiver structure L which is the load of the source.

$$Y_{SL}^A(f) = \frac{v_L^A(f)}{F_H} = \frac{1}{Z_L^A(f) + Z_S(f)}$$

$$Y_{SL}^B(f) = \frac{v_L^B(f)}{F_H} = \frac{1}{Z_L^B(f) + Z_S(f)}$$

(3.36)

Now, the correction function (magnitude and phase) can be expressed by the ratio of the mobilities at the interface between source and receiver structure.

$$H_\Delta(f) = \frac{v_L^B(f)}{v_L^A(f)} = \frac{Z_L^A(f) + Z_S(f)}{Z_L^B(f) + Z_S(f)} = \frac{Y_{SL}^B(f)}{Y_{SL}^A(f)}$$

(3.37)

An advantage of this method is that the coupling between the different directions of space and positions can be considered using dense mobility matrices. They describe the relation between all forces and the resulting velocities. The velocity vector of case B can be predicted if a correction function matrix is multiplied by the velocity vector of case A. The correction function matrix is calculated analogous to the single degree of freedom model in equation 3.37. Instead of the division the mobility matrix of case A is inverted.

$$\mathbf{v}_L^B = \mathbf{H}_\Delta \cdot \mathbf{v}_L^A = \mathbf{Y}_{SL}^B \cdot \left(\mathbf{Y}_{SL}^A\right)^{-1} \cdot \mathbf{v}_L^A$$

(3.38)

The derivation of this formula starts with the relation between the force acting on the receiver structure and the velocity.

$$\mathbf{v}_L^A = \mathbf{Y}_L^A \cdot \mathbf{F}_L^A$$

$$\mathbf{v}_L^B = \mathbf{Y}_L^B \cdot \mathbf{F}_L^B$$

(3.39)

The same force can be calculated from the source velocity and the source and receiver mobility.

$$\mathbf{F}_L^A = \left(\mathbf{Y}_S + \mathbf{Y}_L^A\right)^{-1} \cdot \mathbf{v}_S$$

$$\mathbf{F}_L^B = \left(\mathbf{Y}_S + \mathbf{Y}_L^B\right)^{-1} \cdot \mathbf{v}_S$$

(3.40)

Inserting equation 3.39 in equation 3.40 gives

$$\mathbf{v}_L^A = \mathbf{Y}_L^A \cdot \left(\mathbf{Y}_S + \mathbf{Y}_L^A\right)^{-1} \cdot \mathbf{v}_S$$

$$\mathbf{v}_L^B = \mathbf{Y}_L^B \cdot \left(\mathbf{Y}_S + \mathbf{Y}_L^B\right)^{-1} \cdot \mathbf{v}_S.$$

(3.41)

This equation can be combined to a single equation without the source velocity. After a few rearrangements it is proven that the correction function matrix can be calculated using the mobility matrices at the interface between source and receiver structure of both cases A and B.

$$
\begin{aligned}
\mathbf{v}_L^B &= & \mathbf{Y}_L^B \cdot \left(\mathbf{Y}_S + \mathbf{Y}_L^B\right)^{-1} \cdot \left(\mathbf{Y}_L^A \cdot \left(\mathbf{Y}_S + \mathbf{Y}_L^A\right)^{-1}\right)^{-1} \cdot \mathbf{v}_L^A \\
&= & \mathbf{Y}_L^B \cdot \left(\mathbf{Y}_S + \mathbf{Y}_L^B\right)^{-1} \cdot \left(\mathbf{Y}_S + \mathbf{Y}_L^A\right) \cdot \left(\mathbf{Y}_L^A\right)^{-1} \cdot \mathbf{v}_L^A \\
&= & \left(\left(\mathbf{Y}_S + \mathbf{Y}_L^B\right) \cdot \left(\mathbf{Y}_L^B\right)^{-1}\right)^{-1} \cdot \left(\mathbf{Y}_S + \mathbf{Y}_L^A\right) \cdot \left(\mathbf{Y}_L^A\right)^{-1} \cdot \mathbf{v}_L^A \\
&= & \left(\mathbf{Y}_S \cdot \left(\mathbf{Y}_L^B\right)^{-1} + 1\right)^{-1} \cdot \left(\mathbf{Y}_S + \mathbf{Y}_L^A\right) \cdot \left(\mathbf{Y}_L^A\right)^{-1} \cdot \mathbf{v}_L^A \\
&= & \left(\mathbf{Y}_S \cdot \left(\left(\mathbf{Y}_L^B\right)^{-1} + \left(\mathbf{Y}_S\right)^{-1}\right)\right)^{-1} \cdot \left(\mathbf{Y}_S + \mathbf{Y}_L^A\right) \cdot \left(\mathbf{Y}_L^A\right)^{-1} \cdot \mathbf{v}_L^A \\
&= & \left(\left(\mathbf{Y}_L^B\right)^{-1} + \left(\mathbf{Y}_S\right)^{-1}\right)^{-1} \cdot \left(\mathbf{Y}_S\right)^{-1} \cdot \left(\mathbf{Y}_S \cdot \left(\mathbf{Y}_L^A\right)^{-1} + 1\right) \cdot \mathbf{v}_L^A \\
&- & \left(\left(\mathbf{Y}_L^B\right)^{-1} + \left(\mathbf{Y}_S\right)^{-1}\right)^{-1} \cdot \left(\mathbf{Y}_S\right)^{-1} \cdot \mathbf{Y}_S \cdot \left(\left(\mathbf{Y}_L^A\right)^{-1} + \left(\mathbf{Y}_S\right)^{-1}\right) \cdot \mathbf{v}_L^A \\
&= & \left(\left(\mathbf{Y}_L^B\right)^{-1} + \left(\mathbf{Y}_S\right)^{-1}\right)^{-1} \cdot \left(\left(\mathbf{Y}_L^A\right)^{-1} + \left(\mathbf{Y}_S\right)^{-1}\right) \cdot \mathbf{v}_L^A \\
&= & \left(\left(\mathbf{Y}_S\right)^{-1} + \left(\mathbf{Y}_L^B\right)^{-1}\right)^{-1} \cdot \left(\left(\left(\mathbf{Y}_S\right)^{-1} + \left(\mathbf{Y}_L^A\right)^{-1}\right)^{-1}\right)^{-1} \cdot \mathbf{v}_L^A \\
&= & \mathbf{Y}_{SL}^B \cdot \left(\mathbf{Y}_{SL}^A\right)^{-1} \cdot \mathbf{v}_L^A
\end{aligned}
\tag{3.42}
$$

with

$$
\begin{aligned}
\mathbf{Y}_{SL}^A &= \left(\left(\mathbf{Y}_S\right)^{-1} + \left(\mathbf{Y}_L^A\right)^{-1}\right)^{-1} \\
\mathbf{Y}_{SL}^B &= \left(\left(\mathbf{Y}_S\right)^{-1} + \left(\mathbf{Y}_L^B\right)^{-1}\right)^{-1}.
\end{aligned}
\tag{3.43}
$$

The proposed impedance based method is demonstrated at two abstract miniature vehicle models in the next chapter. The influence of different structure impedances on the acceleration signals is shown and how to compensate this by using correction functions. Furthermore, the effective mount transfer function method and parametric mount models are validated in different examples.

Chapter 4

Results of engine TPA methods

Results and examples of engine transfer path analysis are presented in this chapter. In the first example, structure-borne sound shares are discussed which are calculated by three different methods: indirect force determination (section 2.4), mount attenuation function (section 2.7) method, and the proposed effective mount transfer function method (section 3.1).

In the following, the benefit of using active-side accelerations in the auralization process is explained. In the next section, parametric mount models introduced in section 3.2 are used to determine operational forces induced into the car body through the engine mounts. Estimated mount stiffness data are compared with values given by the manufacturer's data sheet. The coupling within a mount is considered in the following examples dealing with engine mounts and suspension mounts.

A transfer path synthesis based on input signals of a vehicle A or a test rig can be used in a transfer path model of a vehicle B, if correction functions are applied to consider different body inertances. The introduced method described in section 3.6 is demonstrated at two simplified miniature vehicle models.

4.1 Effective mount transfer function method

4.1.1 Comparison of structure-borne TPA methods

[1] As an example, a transfer path analysis of a classic car is performed, where the operational forces are calculated with the three approaches: indirect force determination, mount attenuation without coupling, and effective mount transfer function method considering structure coupling. For all approaches, the structure-borne sound shares of the engine mounts at the position of a receiver microphone at the driver's seat are synthesized using a common set of acoustic transfer functions. The results of a run-up with partial load are shown in Figure 4.1 as spectrograms vs. RPM. The method based on mount attenuation considering no structure coupling overestimates the engine noise. In the next paragraph this is discussed in more detail. With IFD (meaning inverting a dense inertance matrix), there is no overestimation but the influence of the rolling tires and other disturbances are visible. The new approach combines the advantages of both, and the result shows no overestimation and no disturbances.

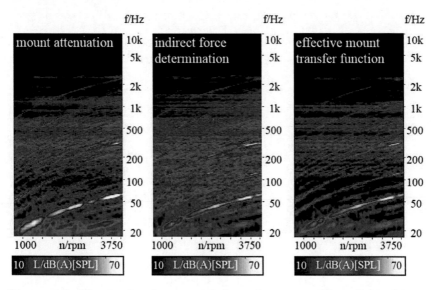

Figure 4.1: *The structure-borne sound share of a run-up with partial load is synthesized using three different approaches.*

---

[1]This part has been previously published in [59].

The quality of a TPA can be evaluated by comparing the synthesis with the measured reference at the receiver position (Figure 4.2 a). The engine of the classic car produces a considerable airborne sound share which must be added to the structure-borne sound synthesis to allow a meaningful comparison with the reference.

The airborne sound share synthesis uses microphones near the engine and airborne sound sensitivities measured with a small loudspeaker. The engine noise synthesis based on mount attenuation including the airborne noise share is quite good, but the main engine order is overestimated (Figure 4.2 b).

If the strong coupling is considered, as in the new approach, the synthesis is better (Figure 4.2 c). Neither synthesis includes the noise caused by the tires rolling on the dynamometer, which can be seen in the measurement between 200 Hz and 1 kHz.

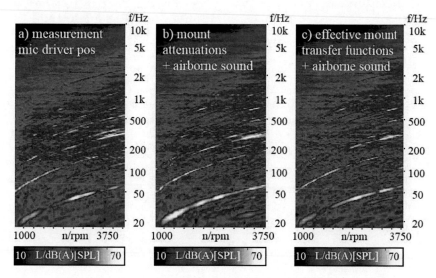

Figure 4.2: *a) The measured interior noise at the driver position; b) engine noise synthesis based on mount attenuation functions plus airborne sound share; c) engine noise synthesis based on effective mount transfer functions plus the same airborne sound share.*

4.1.2 Influence of tire-road excitations on synthesized engine structure-borne noise

[2] In the second example, the influence of the tire-road excitations on the synthesized structure-borne noise share of the engine is studied further. Usually, operational measurements used for an engine TPA are performed on a roller dynamometer. A smooth roller surface is preferred in order to reduce the noise of the rolling tires in the vehicle's interior. Sometimes the measurements must be carried out on a real road due to no dynamometer being available or due to other reasons. In this case the tire-road noise can have a major impact on the synthesized engine noise based on passive side accelerations.

For this example, an engine TPA of a compact van is based on real road measurements with two different road surfaces. A very smooth and a rough road surface were chosen. At the beginning of each measurement the car's speed is a constant 30 km/h. After a short acceleration period the speed is a constant 40 km/h. The structure-borne engine noise is synthesized via two methods: Indirect force determination with matrix inversion using passive-side accelerations, and effective mount transfer functions using active-side accelerations. Figure 4.3 (a) shows the spectrogram of the synthesized engine noise share contributed by the structure-borne sound of the three engine mounts for the scenario of very smooth road surface. The engine noise is quite low except for during the acceleration phase because a high gear was chosen. The synthesis based on effective mount transfer functions contains fewer random noise components and almost only engine orders (Figure 4.3 b). It can be concluded that even on a smooth road surface the passive side accelerations contain tire-road excitations. The same syntheses are calculated for the rough road measurements. It can be assumed that the engine noise will not change significantly if the car is driving on a rough instead of a smooth surface, if all other parameters like payload or temperature do not change. But now the synthesis based on indirect force determination with matrix inversion contains noise components between 50 Hz and 350 Hz which are independent of engine speed (Figure 4.3 c). It is obvious that these components are due to the tire-road noise which is measured on the passive-side of the engine mounts. The active-side accelerations are considerably less disturbed so that the synthesis using effective mount transfer functions (Figure 4.3 d) is very similar to the case of the smooth road surface. The active side accelerations are affected only around 100 Hz. Here the tire-road excitation is too high and the assumption that the active side accelerations contain only engine noise is not completely fulfilled. Even under these extreme circumstances of low engine noise and high tire-road noise, a good structure-borne engine noise synthesis is possible if active-side accelerations are the basis of the TPA/TPS.

---

[2]This part has been previously published in [59].

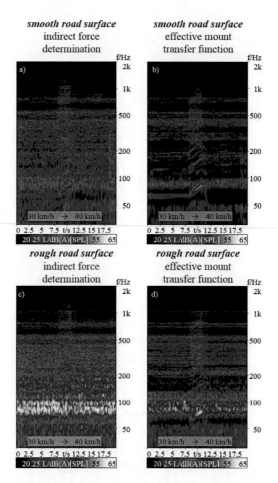

Figure 4.3: *a) The structure-borne engine noise share of a compact van is syn-thesized using passive-side accelerations and Indirect Force Determi-nation with matrix inversion. The synthesis is disturbed by tire-road excitations of the smooth road surface. b) This can be avoided, if the synthesis is based on active-side accelerations and effective mount transfer functions. c) On a rough road the disturbance caused by the tire-road noise is considerably higher. d) The influence on the active-side accelerations is less which leads to a better engine noise synthesis.*

4.2 Parametric mount models

4.2.1 Example Kelvin-Voigt model

For this example[3], the operational forces of a four-cylinder engine are calculated using Kelvin-Voigt models. The parameters damping factor and resilience are determined such that the mount model combined with the structure inertance is a good approximation for the effective mount transfer function. The forces are compared in Figure 4.4. The operating condition is a run-up on a roller dynamometer with full load and 2nd gear engaged. Up to 200 Hz the simple mount model gives a good approximation of the force. The results can be improved if the mass of the mount is considered as described above. Then the forces match up to 1 kHz.

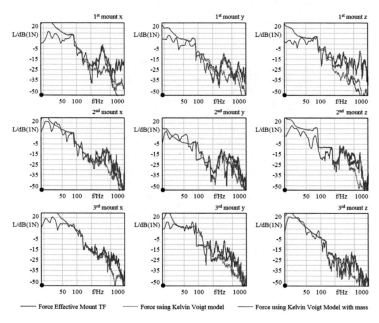

Figure 4.4: *Comparison of operational forces calculated with Effective Mount Transfer Function (blue), Kelvin-Voigt model (gray) and Kelvin-Voigt model with mass (green).*

---

[3]This example has been previously published in [87].

4.2.2 Stiffness parameter estimation

In the next example[4], for each mount of a mid-size car three independent Kelvin-Voigt models are used. One stiffness and one damping parameter characterize each direction. These parameters are determined such that the mount model impedance is a good approximation of the mount impedance calculated from effective mount transfer function and the structure inertance according to equation 3.13. The operating condition is a run-up in the 3rd gear under full load measured on a roller dynamometer.

In this case, static stiffness data of the mounts was available. The estimated values are compared to the values given by the manufacturer's data sheet in Figure 4.5. There is a good agreement between both versions, but the deviation of the engine mount stiffness in y-direction and the pendulum support stiffness in x-direction are higher. Nevertheless, the data sheet values are only typical values, which are determined on a test rig using identically constructed mounts and not the actual mounts of the test car. On the contrary, the estimated values consider the real assembly situation in the car, which is a big benefit.The operational forces, which are induced through the engine mounts into the car body, are shown in Figure 4.6 as averaged spectra. Despite a relatively simple mount model there is quite a good agreement to the result of the indirect force determination based on inverted structure inertance matrices and passive-side accelerations.

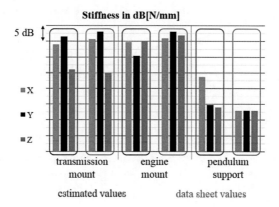

Figure 4.5: *Comparison of estimated stiffness data with the values given by the manufacturer's data sheet*

---

[4]This example has been previously published in [97].

4.2.3 Engine mounts 3D Kelvin-Voigt model

[5] The next step is to use models that include the directional coupling for the same example as described above.

An increased quality of the forces estimated by the model is expected if the directional coupling is considered by using three Kelvin-Voigt models per direction for each mount. In fact, the resulting operational forces show a better accordance with the indirectly determined forces (Figure 4.6).

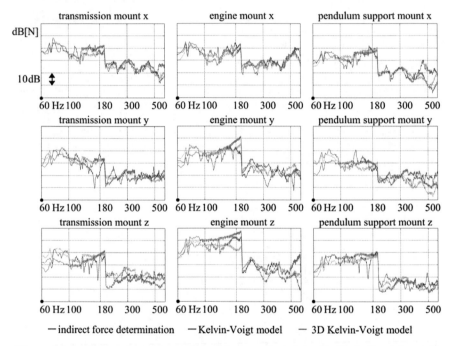

Figure 4.6: *Comparison of operational forces calculated with the indirect force determination (black), one Kelvin-Voigt model per direction (blue) and three Kelvin-Voigt models per direction (green)*

---

[5]This example has been previously published in [97].

4.2.4 Suspension mounts 3D Kelvin-Voigt model

[6] The proposed method is not limited to engine mounts. It can be applied to other mounts or bushings in a car, too. Another example deals with the mounts of a triangle transverse control arm (TTCA), which is an important part of the suspension. For validation purposes, the TTCA is assembled on a special test rig with force sensors (Figure 4.7) [58]. This allows for an evaluation of the forces calculated using the mount models by comparing them to the directly measured forces. Unfortunately, this is almost impossible in a real situation where the TTCA is installed in the car. The dimensions of a calibrated triaxial force sensor are too high to place them between the control arm and sub-frame of the car.

The TTCA has two rubber mounts, one at the front position and one at the rear position. Mount adaptors were crafted to connect the TTCA to calibrated triaxial force sensors. This set-up is placed on a base plate and a concrete foundation. A shaker is the excitation source and replaces the wheel. The active-side accelerations are measured on the TTCA, the forces which are induced into the base plate are recorded simultaneously. Each mount is modeled by nine damping and nine stiffness parameters according to Figure 3.11. In this case, the parameters are optimized so that the forces calculated by the effective mount transfer functions from the model (equation 3.28) best fit the directly measured forces. In other cases, like in a car, the indirectly determined forces would be used instead.

The directly measured and calculated forces based on the mount models during operation of the shaker are shown in Figure 4.8 as averaged spectra. There is a good accordance between the forces of both methods, which shows that the chosen model describes the system behavior very well.

---

[6]This example has been previously published in [97].

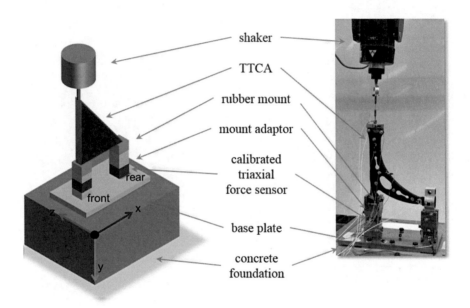

Figure 4.7: *Test rig for transverse control arm with triaxial force sensors [58].*

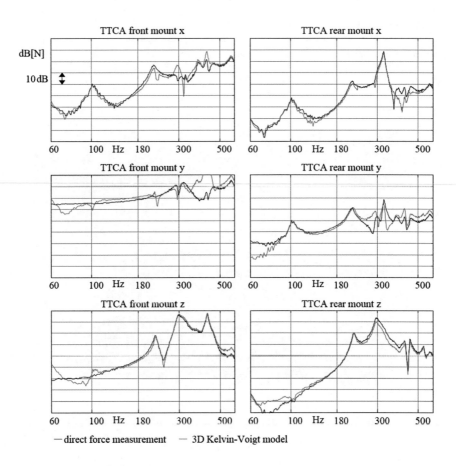

Figure 4.8: *Comparison of operational forces measured with triaxial force sensors (black) and calculated by mount models containing three Kelvin-Voigt models per direction (green).*

4.3 Transfer path synthesis based on engine test rig data

Input signals of an engine measured on a test rig or at a vehicle A can be used for a transfer path synthesis based on a TPA model of a different vehicle B if the influence of the different load impedances is considered. Frequency dependent correction functions can be determined from impact measurements of the assembled systems, as it is shown in chapter 3.6, to adapt the acceleration signals. The feasibility of this method is validated using two simplified miniature vehicle models.

Figure 4.9: *vehicle simulator A*

The miniature vehicle models A and B simulate a car with a reduced complexity (Figure 4.9 and 4.10). Each vehicle simulator consists of a body, a small cabin, and an engine which is a shaker with a mass connected via a mount to the body. The shaker is a structure-borne noise source. The vehicle simulators have different bodies, but the same engine is first mounted on A and then on B. A triaxial accelerometer is placed on the shaker mass (active side) and a second one on the body next to the mount (passive side). For the operational measurement the shaker plays back a sweep signal from 40 Hz to 2 kHz via the sound card of a laptop which is connected to a small amplifier. The voltage at the shaker terminals is also recorded as a reference signal.

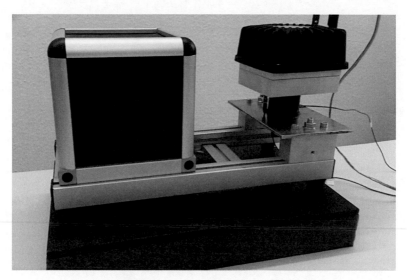

Figure 4.10: *vehicle simulator B*

In Figure 4.11 the averaged spectra of the acceleration signals in z-direction during shaker excitation are shown. There is a difference between cases A and B, although the shaker voltage is identical, because of the different body impedances. The active-side accelerations are different between 70 Hz and 140 Hz about 10 dB and the passive-side accelerations up to 40 dB at 800 Hz due to a resonance in case B. The task is to predict the accelerations which would be measured in case B from the accelerations of case A.

At first, only the z-direction is considered. The correction function $H_\Delta(f)$ describing the difference between both versions can be calculated, according to equation 3.37, from the inertance frequency response functions, which are measured with an impact hammer at the active side and the passive side, respectively, for both cases A and B (Figure 4.12). The amplitude and the phase of the correction functions are shown in Figure 4.13 and in Figure 4.14. There is a very good accordance with the actual measured difference between case A and B. At most frequencies the deviation is less than 2 dB while at some frequencies the deviation is between 3 and 5 dB. Here, it is sufficient to neglect the coupling of the directions because there is only one major direction of excitation.

The coupling between the different directions must be considered if the other directions x and y are analyzed, which are dominated by the excitation in z-direction. This is achieved using a correction function matrix calculated from dense inertance

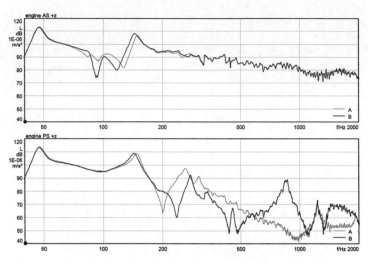

Figure 4.11: *The averaged spectra of the acceleration signals during the shaker excitation are not equal for cases A and B due to different body impedances.*

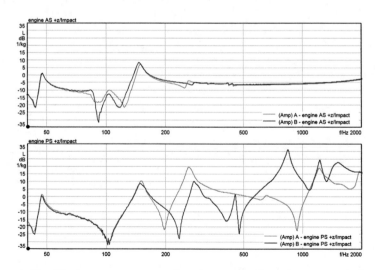

Figure 4.12: *The inertances measured at the active side and passive side for cases A and B are shown.*

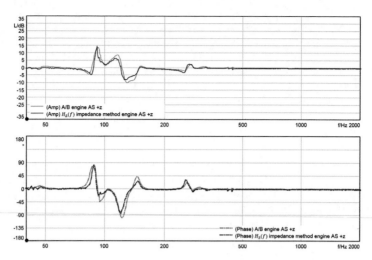

Figure 4.13: *The frequency dependent ratio between the active-side accelerations in z-direction of the case A and B and the correction function $H\Delta(f)$ calculated from the inertances shows a very good accordance.*

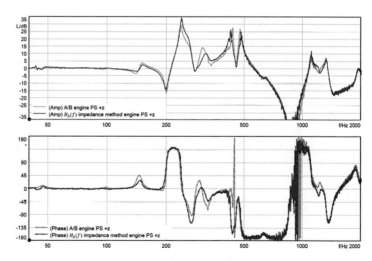

Figure 4.14: *The frequency dependent ratio between the passive-side acceleration in z-direction of the case A and B and the correction function $H\Delta(f)$ calculated from the inertances shows a very good accordance too.*

matrices according to equation 3.38. The impact measurements are performed in all three directions. Now, the accelerations of case A are predicted using the accelerations of case B and the correction function matrix. The results are shown in Figure 4.15 and Figure 4.16. The acceleration signals are related to the shaker voltage. Without correction function the differences between cases A (green) and B (red) are intolerable high with a maximum deviation of 40 dB. After applying the determined correction functions the predicted accelerations (blue) are very similar to the case A (green). The deviation is less than 3 dB, except for the y-direction below 200 Hz, where the difference is up to 10 dB but here the amplitude is lower compared to the other directions.

The proposed method of determining correction functions based on inertances has been successfully demonstrated at this example. A different car body impedance leads to different accelerations if the same source is attached. The passive-side acceleration signals are more affected than the active-side acceleration signals. The difference can be compensated by applying a frequency dependent correction function (matrix).

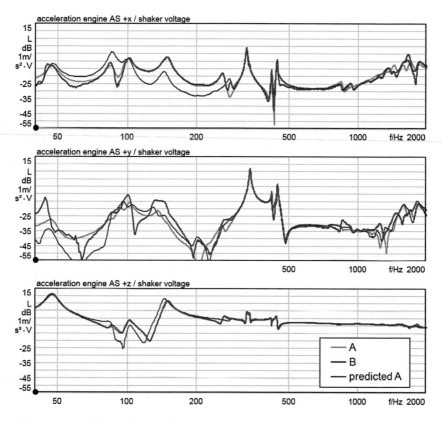

Figure 4.15: *The operational active-side accelerations of case A can be predicted from the accelerations of case B based on inertance measurements.*

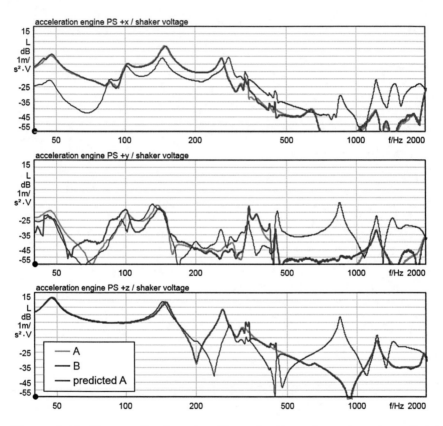

Figure 4.16: *The operational passive-side accelerations of case A can be predicted from the accelerations of case B based on inertance measurements.*

4.4 Summary on engine TPA

In chapter 3 and chapter 4 methods and examples of transfer path analysis for engine applications are presented. The key aspects of the scientific contribution are summarized in the following enumeration.

- Effective mount transfer function method

    - The proposed effective mount transfer function method is a transfer path analysis approach for structure-borne sound paths which includes mount characteristics.

    - The effective mount transfer functions are determined in-situ, that means, engine or mounts must not be removed during the measurements. This saves time and prevents the system being changed due to disassembling and reassembling. Measuring mount characteristics on a special test rig are omitted. The challenges of crafting adaptors and setting correct preload are avoided.

    - The additional measurement effort compared to standard inverse force identification with matrix inversion is marginal. Only some additional acceleration sensors placed on the engine next to the mounts are needed.

    - With the effective mount transfer function no predictions can be made for the case that the source is mounted on another receiver structure with different inertances. Then mount parameters, e.g. two-port impedances, are necessary. In the case of a two-port with only a series circuit and no shunt circuit the two-port impedances can be calculated from the EMTF and the structure inertance.

    - The coupling of the receiver structure is considered.

    - The auralization is performed by filtering active-side accelerations in the time domain. The quality of the synthesis is increased because active-side accelerations are less disturbed by other sources than passive-side accelerations, which are usually used in the inverse force identification method. The synthesized structure-borne engine shares based on active-side accelerations are much clearer and less noisy.

- Parameter mount models

    - The effective mount transfer functions are a very good basis for estimating parameters of mount models. Mount models with physical parameters help to predict how to modify a mount.

- Simple models with only a few parameters have the advantage of being more robust against overfitting than complicated models with many parameters.

- Kelvin-Voigt models already achieve good results. The parameters, stiffness and resilience are estimated using in-situ measurements only. Special test rigs are not necessary.

- The prediction quality can be increased if a directional coupling within a mount is considered. That means an excitation in one direction has an effect on the other directions too.

- If three Kelvin-Voigt models per direction are used, the estimated forces can be improved.

- The method has been successfully demonstrated at engine mounts and mounts of a triangle transverse control arm, which is an important part of the chassis.

- Avoiding auralization artifacts

  - An auralization offers the possibility that the engineer, a decision-maker or a customer can experience the results of a transfer path synthesis by listening. It is imperative that the auralization does not contain artifacts, which distract the listener from the proper sound and lead to another perception.

  - For the widely used inverse force identification method with matrix inversion an approach is proposed that eliminates tonal components produced by an inversion of ill-conditioned matrices. These tonal components are generated by incorrect, narrow peaks in the transfer functions.

  - The disturbing peaks in the apparent mass transfer functions are detected by considering high values in the matrix condition and the wrong values are replaced by interpolation.

- Transfer path synthesis based on engine test rig data

  - Input signals of an engine measured on a test rig or at a vehicle A can be used for a transfer path synthesis based on a TPA model of a different vehicle B, if the influence of the different load impedances is considered.

  - A method is proposed which determines frequency dependent correction functions from impact measurements of the assembled systems to adapt the acceleration signals.

  - The feasibility of this method has been successfully demonstrated using an example with two miniature vehicle models.

The presented methods help to find dominating sources and their transfer paths. This information is very useful for developing countermeasures that improve noise issues and lead to quieter engines. In the last few decades the engine noise has been reduced continuously, so that the noise generated from rolling tires attracts more and more attention for sound improvement. With regard to the electrification of the powertrain, tire-road noise will become even more important for sound quality and NVH comfort because of lower masking by the electric motor noise. The presented methods would not be the best choice for tire-road noise analysis where other approaches are more promising. The following chapters deal with methods and examples for chassis TPA and the auralization of tire-road and wind noise.

Chapter 5

Chassis TPA - Tire-road noise analysis

Figure 5.1: *Tire-road noise attracts more and more attention for sound improvement.*

[1] In vehicle acoustics the chassis is, besides powertrain, another important area for the application of transfer path analysis. TPA can help the engineer to find a trade-off between vehicle dynamics and comfortable tire-road noise. In this chapter a new approach is presented to synthesize the tire-road interior noise under dynamic driving conditions without disturbing engine or wind noise shares. The measurement setup is explained and a cross-talk cancellation algorithm is presented to eliminate unwanted sound shares of the engine in the tire-road noise synthesis.

---

[1]This chapter is a revised version taken from [98].

5.1 Auralization of tire-road noise

For a subjective sound quality evaluation of tire-road noise, authentic sound examples are necessary. Different tires or vehicles in a benchmark can be compared to show weak spots or to illustrate improvements in the development process. Interior noise recordings on a roller dynamometer have the advantage of measuring only tire-road excitations without aerodynamic and engine noise, while the rollers drive the tires with the engine switched off. The dynamometers are usually designed for powertrain investigations and their small roller circumference and smooth surface (see Figure 5.2) lead to an unwanted artificial and periodic tire-road noise. Therefore, it is highly recommended that measurements are taken on a real road and not on a dynamometer. But then there is the challenging task of separating the interior noise components generated by rolling tires, aerodynamic flow and powertrain in on-road measurements, so that these noise shares can be judged individually and independently. For sound design or in an acoustic driving simulator it can be useful to combine these components after manipulating them separately.

The engine noise can be simply eliminated by coasting down with the engine switched off after acceleration to the requested maximum speed. Then the task is to separate the measured interior noise into tire-road and aerodynamic noise [99, 100]. Coast-down measurements without the engine running have no practical relevance to customers' driving experience, thus they are unsuitable for listening tests. Dynamic driving conditions with the engine running are more related to customers' experiences and they can influence tire-road noise, e.g., during high acceleration. In these cases, the engine, as an additional noise source, increases the difficulty of separating the interior noise into its components.

Figure 5.2: *Roller dynamometers are usually designed for powertrain investigations and their surface (left side) is too smooth compared to a real road (right side).*

5.2 Data acquisition

The separation of the interior noise into the sound shares of the different sources requires a special measurement setup which is shown in Figure 5.3. During operating conditions, the vehicle interior noise is usually recorded by an artificial head placed on the front passenger's seat, or by the driver wearing a binaural microphone headset. A binaural recording preserves auditory spaciousness and allows for an aurally adequate reproduction using head phones. Of course, a microphone can also be used at an ear position. In addition, signals are measured close to the sound sources. The tire-road noise is generated by the contact of the rolling tires with the road surface. Thus the airborne radiation of each tire is recorded with one microphone placed at the inlet and one at the outlet of a tire. Another microphone can be mounted in the wheelhouse above the tire according to requirements.

Figure 5.3: *Measurement setup for tire-road noise analysis*

The structure-borne part is acquired with a three-dimensional acceleration sensor on each wheel carrier. Source signals of the aerodynamic flow cannot be recorded because there is no clearly-defined source location. The whole chassis and windows are excited. In the case of a coast-down measurement the setup is complete. During dynamic driving conditions, vibrations on the engine are recorded with accelerometers applied close to the mounts. All input signals must be measured synchronously, so that a mobile multi-channel measurement system is required.

5.3 Analysis of coast-down measurements

For coast-down measurements there are at least twenty input signals, for each tire two microphone signals and three acceleration signals, in the MIMO model shown in Figure 5.4. The receiver signals $y_j(t)$ correspond to the binaural head with i=1 for its left and $i = 2$ for its right channel. The aerodynamic noise share of the receiver is modeled as an additive component $n_i(t)$. It is uncorrelated with the tire-road excitations due to their different physical origins. The transfer functions can be calculated using OTPA (see section 2.8) and the tire-road noise is auralized by filtering the input signals $x_j(t)$ with the appropriate impulse responses. It is also possible to synthesize the airborne or structure-borne noise shares separately by summing the microphone signals or the acceleration signals, respectively. The noise shares of the different tires or axles are calculated by considering only the corresponding transfer paths. The noise share caused by the aerodynamic flow, the wind noise, corresponds to the difference of the measurement $y_i(t)$ and the tire-road noise synthesis $t_i(t)$. Therefore, this method allows separating the interior noise into tire-road and wind noise shares.

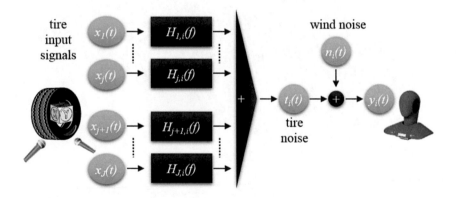

Figure 5.4: *Multiple-Input-Multi-Output model for tire-road noise analysis of coast-down measurements*

In the past the separation of tire-road and wind noise was performed using multiple coherence filtering [101]. Here, the interior noise was filtered according to the Wiener-Filter theorem [102] to achieve an estimation of the tire-road and wind noise. The OTPA method has many advantages compared to the multiple coherence filtering [99, 103].

5.4 Tire-road noise under dynamic driving conditions

5.4.1 Motivation

Under dynamic driving conditions, e.g., full-load acceleration, the engine is necessarily running and is a further source for the interior noise. In principle, this is not an obstacle for OTPA because engine noise is also uncorrelated to tire-road noise. Then the signal $n_i(t)$ contains both wind and engine noise (Figure 5.5). The tire-road noise could still be calculated if there were no crosstalk of the engine to the tire input signals. The engine noise is not only noticeable in the vehicle interior, but is also recorded by the microphones at the tires. Furthermore, the engine introduces forces into the chassis, which also cause vibrations at the wheel hub. It is no longer possible to measure clear tire input signals because of the crosstalk. If the method as described above for the coast-down is applied, this would lead to engine shares in the tire-road noise synthesis. The OTPA transfer functions would be incorrect and the filtered tire input signals and therewith the tire-road noise auralization would contain unwanted engine shares. Pulling the car is not an alternative because the second vehicle's noise would affect the input and receiver signals as well. The load conditions would also be different. For this reason, a cross-talk cancellation (CTC) algorithm must be carried out to eliminate the engine noise shares in the tire input signals.

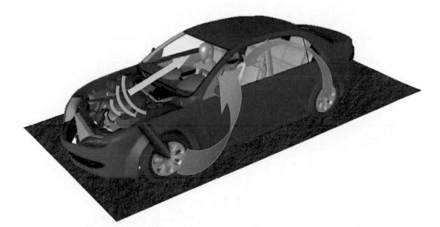

Figure 5.5: *There is a crosstalk of the engine to the tire input signals under dynamic driving conditions.*

5.4.2 Cross-talk cancellation (CTC)

The model of tire-road noise analysis with CTC of engine noise under dynamic driving maneuvers is shown in Figure 5.6. The approach consists of two steps. First, additional signals $z_l(t)$ which are measured on the engine are used to calculate new engine-free tire input and receiver signals. The second step is equivalent to the case of a coast-down without a running engine as mentioned before. Now the transfer functions are calculated correctly from these new signals and the tire-road noise no longer contains engine noise.

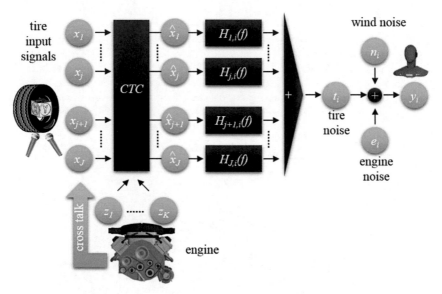

Figure 5.6: *Cross-Talk Cancellation (CTC) for tire-road noise analysis under dynamic driving condition*

Under dynamic driving conditions it is not possible to measure tire input signals without engine shares, but signals characterizing only the engine noise can be recorded. Because of soft mounts and high engine mass the influence of the tire-road excitation on the motor vibrations is negligible.

This can be exploited in the CTC process. Of course the engine noise shares in the tire input signals are correlated with the vibrations on the receiver because they have the same source. Operational transfer path analysis is used to calculate these correlated shares $x_j^{engine}(t)$, shown in Figure 5.7. The acceleration signals on the engine are the input and the tire signals are the output signals of the OTPA model.

After determining the transfer functions, the engine acceleration signals are filtered with the associated impulse responses and summed. Finally, this sum is subtracted from the measured tire input signal $x_j(t)$ in the time domain. The result is a new tire input signal $\hat{x}_j(t)$ without engine noise shares.

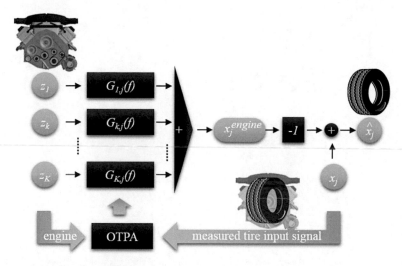

Figure 5.7: *Operational transfer path analysis is used for cross-talk cancellation for tire input signals.*

The airborne radiation of the engine is not considered for the CTC because engine microphone signals contain also tire-road noise due to crosstalk. So they cannot be used in this CTC approach, otherwise tire-road noise would be cancelled. Calculating the correlated engine noise shares is the aim. Therefore, physically correct transfer functions are not necessary. OTPA only evaluates correlation between signals. That is utilized here because the structure-borne and airborne sounds of an engine are highly correlated. In particular, this is valid for the engine orders. Parts of the engine airborne radiations which are not correlated with the engine vibrations cannot be eliminated using this method. The important engine orders are not concerned. Maybe some sound components at higher frequencies are uncorrelated, because they have a different generation mechanism like aero-acoustical sound sources such as the radiator fan or a belt transmission. But at higher frequencies crosstalk and transmission into the vehicle's interior is much lower. In the examples below it can be seen that the uncorrelated part has no relevance and can be neglected. The frequency range of the CTC is limited by the cut-off frequency of the accelerometer on the engine. In most cases the dominant engine orders are below this frequency.

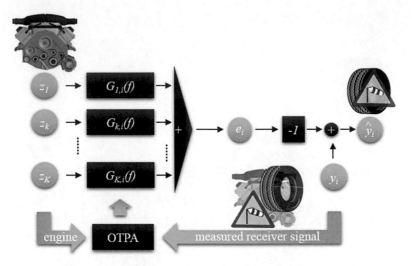

Figure 5.8: *Cross-Talk Cancellation for receiver signals*

In the same way, the engine noise share in the receiver signals is removed by synthesizing the engine noise $e_i(t)$ at the receiver locations using $z_l(t)$ and $y_i(t)$ and subsequent subtraction from the measured signals $y_i(t)$ (Figure 5.8). In the second step a further OTPA is performed with the new tire input and output signals. Now correct transfer functions are calculated and the modified input signals which do not contain engine noise anymore are filtered and summed (Figure 5.9).

The wind noise can be determined by subtracting the tire-road and engine noise from the measured interior noise signals. It must be noted that the calculated wind noise would contain any disturbances included in the interior noise recording that is uncorrelated to engine and tire-road excitations. This could be the squeak and rattle of the car interior or noise made by the driver.

The new approach determines the tire-road noise of on-road measurements with the engine running but without disturbing engine noise shares in the auralization. The crosstalk from the engine to the tire input signals is shown at an example in the next chapter. If the standard method is applied, the calculated tire-road noise contains unwanted engine noise. With the new method the correct tire-road and wind noise shares are synthesized during a run-up under full load.

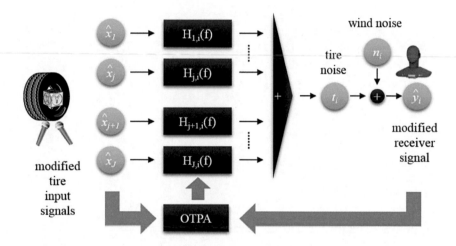

Figure 5.9: *OTPA after Cross-Talk Cancellation with modified tire input and receiver signals*

Chapter 6

Results of chassis TPA methods

This chapter contains the results of the chassis TPA examples. First, the tire-road noise of a coast-down measurement is synthesized. In the next example, the combustion engine is running to synthesize the tire-road noise under dynamic driving conditions according to the approach presented in section 5.4. The proposed crosstalk cancellation is also demonstrated in the case of an electrical vehicle. The operational transfer path analysis (section 2.8) is compared to the conventional TPA (section 2.4) in view of chassis applications. Finally, a comparison of direct force measurement with indirect force determination for tire-road noise analysis is performed.

6.1 Sound shares of the tire-road noise

6.1.1 Coast-down

[1] In an example the sound shares of the tire-road noise during a coast-down of a luxury-class car are calculated using operational transfer path analysis according to section 5.3. The vehicle interior noise was recorded with an artificial head on the front passenger seat while the engine was switched off. The speed range goes from 120 to 100 km/h.

All OTPA results of the tire-road noise contributions are depicted as averaged spectra in Figure 6.1 (left receiver channel) and Figure 6.2 (right receiver channel); each row of the 2D-plots represents one spectrum. The tire-road noise is divided into the contributions of front and rear axle and each tire. This is continued for all of the airborne and structure-borne noise shares. Additionally, spectra of the measurement as well as of the wind noise estimation are shown in the first two rows.

---

[1]This part is a revised version taken from [103].

artificial head, front passenger seat, left channel

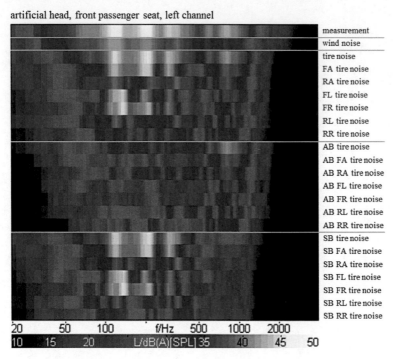

Figure 6.1: *Contributions of tire-road noise (left receiver channel) using OTPA for a luxury-class car (120 to 100 km/h). Abbreviations: SB: structure-borne sound, AB: airborne sound, FA: front axle; RA: rear axle, FL: front left, FR: front right, RL: rear left, RR: rear right.*

It is clearly visible that the tire-road noise is based on the structure-borne contribution below 500 Hz and on airborne sound above this frequency. Furthermore, the front axle, especially the front right tire, contributes more than the rear axle. In this example the receiver was placed on the front passenger seat.

The right ear-signal of the artificial head is louder than the left ear-signal because of the lesser distance of the right ear to the window. In this case, the spatial perception of noise shares at the left and right side is audible in the synthesized signal of the left and right tires as well as in the vehicle interior measurement.

artificial head, front passenger seat, right channel

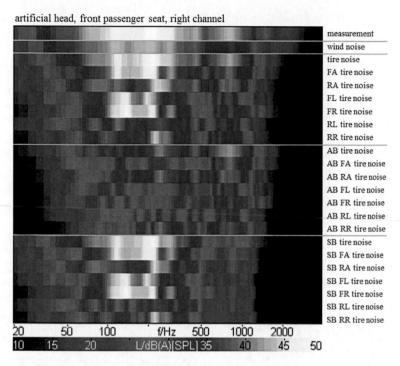

Figure 6.2: *Contributions of tire-road noise (right receiver channel) using OTPA for a luxury-class car (120 to 100 km/h). The receiver is an artificial head (right channel) at the front passenger seat. Abbreviations: SB: structure-borne sound, AB: airborne sound, FA: front axle; RA: rear axle, FL: front left, FR: front right, RL: rear left, RR: rear right.*

6.1.2 Dynamic driving conditions

[2] This example covers an acceleration driving maneuver from 40 km/h to 110 km/h of a compact van on a test track under full load. The engine speed is from 2400 rpm up to 4200 rpm. For a comparison a coast-down was measured with mirror-image speed range on the same road with engine idling. For security reasons, namely to enable brake force booster and airbags, the engine was not switched off. The results of CTC and tire-road noise syntheses are presented.

CTC applied on tire signals

First of all, the crosstalk in the tire input signals should be evaluated. In Figure 6.3 a) and b) the spectrograms of the acceleration on the front left wheel hub in z-direction are compared in the cases of coast-down and run-up.

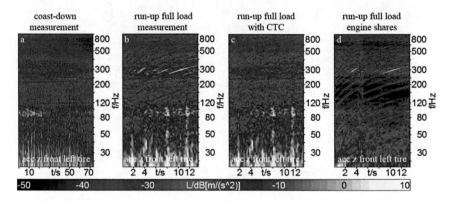

Figure 6.3: *FFT vs. time of the acceleration on the front left wheel carrier in z-direction. From left to right: a) measurement during a coast-down with engine idling; b) measurement during a run-up on the same road under full load; c) result after CTC; d) calculated engine shares*

The frequency range from 20 Hz to 800 Hz covers the relevant area where the structure-borne sound share usually dominates the vehicle interior tire-road noise [103]. The tire cavity resonance around 235 Hz is visible in both figures. The duration of the coast-down which is limited by rolling and wind resistance is five times longer than the run-up. The crosstalk of the powertrain is clearly distinctive during run-up in the form of engine orders because the engine introduces forces

---

[2]This example has been previously published in [98].

into the car body, the suspension and the wheel hub. There is a resonance band between 250 Hz and 350 Hz where tire-road excitation as well as cross-talk are high. The next diagram Figure 6.3 c) contains the result after CTC. The engine orders can be removed with the presented approach and the spectrogram is similar to the coast-down. Figure 6.3 d) shows the calculated engine noise shares which are used to cancel the engine noise. It is the part of the wheel hub acceleration which is correlated to the engine vibrations.

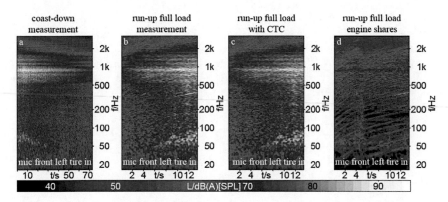

Figure 6.4: *A-weighted spectrograms of the microphone signal positioned at the inlet of the front left tire are shown. From left to right: a) measurement during a coast-down with engine idling; b) measurement during a run-up on the same road under full load; c) result after CTC; d) calculated engine shares*

Figure 6.4 displays the same diagrams for an airborne sound input signal. The spectrogram of the microphone at the inlet of the front left tire has its maximum at higher frequencies, in particular around 1 kHz: The crosstalk of the engine is less visible compared to the wheel hub acceleration, but still audible. With increasing frequency, the influence of the engine decreases. In Figure 6.4 c) the CTC has removed the orders and the result is similar to the coast-down measurement except for the shorter duration and the mirror-image speed range. The engine noise shares are shown in the last diagram.

Tire-road noise syntheses

The crosstalk of the engine to the tire input signals has been demonstrated. Now, the tire-road noise syntheses of this example are calculated with and without CTC. The receiver is an artificial head on the front passenger's seat. A-weighted spectrograms of its left ear are shown in Figure 6.5. The measured receiver signal during the run-up is shown on the top left. The engine orders, the tire cavity resonance and the tire excitation around 1 kHz can again be found. As expected, the tire-road noise synthesis also contains engine noise, if no cross-talk cancellation is carried out. The difference between the structure-borne sound (SB) and airborne sound (AB) shares proves that the crosstalk mainly takes place in the structure, but that the airborne sound shares are also affected (Figure 6.5 c) and d)). The CTC can remove the engine noise shares in the signals so the tire-road noise synthesis no longer contains engine noise. It can be seen that up to 500 Hz structure-borne sound, and above airborne sound, are the dominant sources of the tire-road noise in the vehicle's interior. Although only accelerations on the engine are used for CTC, significant engine orders can be eliminated. Some high frequency orders with lower levels cannot be removed completely, but are usually masked by the tire-road noise. The engine noise shares inside the vehicle can also be synthesized (Figure 6.5 e)). An overview of the individual interior noise sound shares from tire-road contact, engine and wind flow are shown in Figure 6.6.

Comparison of run-up with coast-down (3ʳᵈ gear)

In the next example a run-up under partial load is compared to a coast-down with third gear engaged. The car is accelerated from 20 km/h to 110 km/h in 25 seconds and then coasts down back to 20 km/h. The coast-down phase takes almost twice as long as the acceleration phase, i.e., approximately 50 seconds. The interior noise is separated into its components of tire-road, engine and wind noise. The spectrograms of the artificial head's left channel are shown in Figure 6.7. The area around 220 Hz and 1 kHz is caused by the rolling tires. Relevant engine orders are spread from 50 Hz to 500 Hz. Wind noise is the main source for the interior noise above 2 kHz.

The tire-road noise synthesis in Figure 6.7 b) contains no unwanted engine orders. Here the CTC should be performed separately for run-up and coast-down (3ʳᵈ gear) because the mount characteristics differ for the two different load conditions of the engine. That means that the relationship between the accelerations on the engine side which are used for CTC and the forces induced to the car body are changed. Therewith, the crosstalk of the engine to the wheel hub is different. There is no noticeable difference between run-up and coast-down (3ʳᵈ gear) in the tire-road noise. As expected, the engine noise is louder during run-up. High engine speed causes dominant orders. The wind noise still contains a few slight orders which CTC was unable to remove completely.

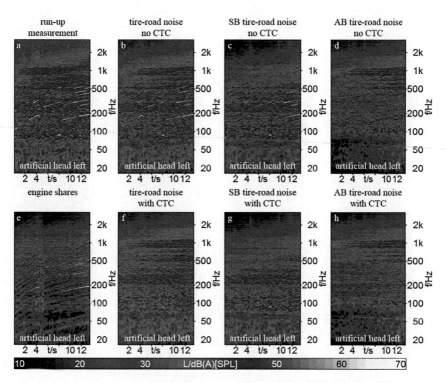

Figure 6.5: *a) A-weighted spectrogram of the artificial head's left ear during the run-up measurement; b)-d) result of the tire-road noise synthesis and the airborne sound (AB) and structure-borne sound (SB) noise shares without a CTC in the first row; f)-h) results with CTC in the second row; e) diagram on the bottom left shows the engine shares*

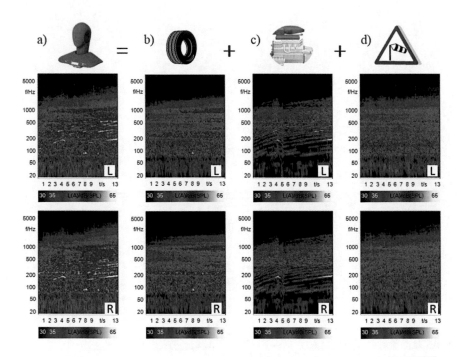

Figure 6.6: *The interior noise measurement (a) can be separated into noise shares of tire-road contact (b), engine (c) and wind flow (d). Spectrograms are shown for the left (L) and right ear (R) of a binaural head placed on the front passenger's seat.*

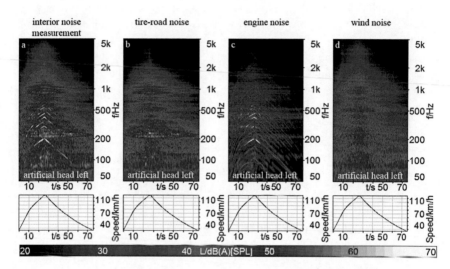

Figure 6.7: *a) The spectrogram of the interior noise during a run-up from 20 km/h to 110 km/h under partial load with a subsequent coast-down (3ʳᵈ gear) to 20 km/h is shown; b) tire-road noise synthesis with CTC; c) synthesized engine share; d) wind noise. The car's speed is plotted below.*

Electric car

As an example, further investigations were made to see if the method of eliminating engine noise is transferable to electric cars. Although an electric motor causes significantly lower noise than a combustion engine, it can still be noticeable in the vehicle's interior, at least under full load. There is also a crosstalk between engine and tire input signals. A further difference between the two powertrain technologies is the sound characteristic. An electric car produces higher orders and may sound like a streetcar or light-rail car. Additionally, the inverter generates high frequency components. The question is if the engine noise can be sufficiently reduced with the different boundary conditions such as higher frequency range, different mounting and lower engine mass.

An electric sub-compact car was set up with two microphones at each tire and a microphone at the receiver position on the front passenger's seat. Two acceleration sensors were placed on the electric motor and the gear box for the CTC according to Figure 5.7. Here, a further accelerometer placed near the inverter cannot be used because it contains explicit tire-road noise components like the tire cavity resonance around 250 Hz. The structure-borne isolation of the converter housing is apparently too low. If this sensor signal were to be used, a part of the tire-noise would be cancelled.

In Figure 6.8 the spectrogram of the interior noise is shown for the operating condition of acceleration under full load from 50 km/h to 70 km/h. The inverter noise, tonal components around 7 kHz, is highlighted. In comparison to combustion engines, which would have a main order between 100 Hz and 200 Hz, the electrical motor transmitted a dominant order between 500 Hz and 700 Hz into the vehicle interior. Other orders are masked by tire-road noise. The tire cavity resonance is distinctive at 250 Hz. With the help of the CTC approach the motor and inverter noise can be reduced significantly (Figure 6.8 b)). The tire-road and wind noise remain. Due to high frequency components generated by the inverter, accelerometer signals are sampled at the same rate as the microphones.

The crosstalk of the powertrain to the tire input signals is especially relevant to the microphones of the rear axle where the electric motor and the inverter are mounted. The spectrogram of the microphone placed at the inlet of the rear left tire is shown in Figure 6.8 c. As expected, the inverter noise is clearly visible. It can be reduced considerably by CTC. A small share which is not masked by other sources remains.

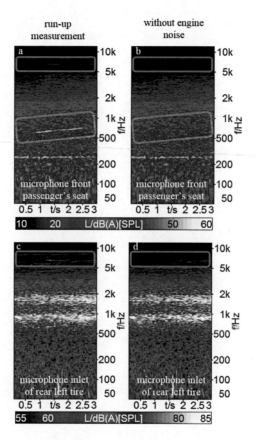

Figure 6.8: *a) A-weighted spectrogram of a measured interior noise during a run-up with an electric car; b) interior noise after removing electrical engine noise; c) measurement of tire input signal; d) result after CTC*

6.2 Comparison of OTPA and conventional chassis TPA

The structure-borne sound share of the tire-road noise and the contribution of the axles or tires can be calculated using operational transfer path analysis (section 2.8) and also conventional transfer path analysis (indirect force determination, see section 2.4).

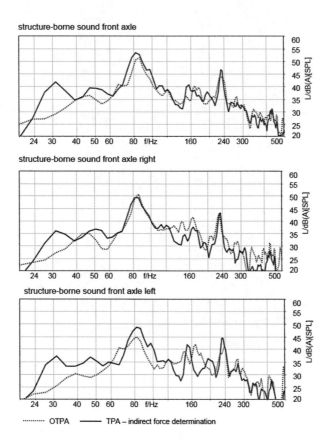

Figure 6.9: *Structure-borne sound share of the tire-road noise on a medium rough road surface at a speed of 30 km/h are calculated with OTPA and conventional TPA using indirect force determination.*

The advantages of OTPA are that it requires less effort but produces very good results. But this method is not suited for highly coherent input signals where the estimated path contribution would be wrong [89, 57]. The signals measured at the different tires and axles can be assumed to have low coherence which fulfills a requirement of OTPA. In contrast, the signals measured on the individual components of the chassis like control arm or strut are highly coherent. That means OTPA cannot be used for a breakdown into more detailed paths and a conventional TPA with indirect force determination is inevitable.

Both approaches will be compared in an example of a compact class car at a constant speed of 30 km/h on a medium rough road surface. The OTPA is based on accelerations measured on the wheel hub. The indirect force determination uses other sensors which are placed on the subframe next to the mounts of the transverse control arms. Further sensors are mounted at the strut bearing and the radius road. Additional positions are used for the purpose of overdetermination.

The structure-borne sound share of the front axles is calculated by adding all paths of the front axle. The shares of the left or right front axle are determined by adding only the corresponding paths. The structure-borne syntheses of OTPA and TPA are shown in Figure 6.9. The results of both approaches are very similar. An explanation for the deviation below 60 Hz could be the fact that it is difficult to measure inertances at low frequencies where the coherence of impact hammer measurements can be low. This example shows, that conventional TPA can be used to calculate the sound shares of the different tires, too. On the other hand, OTPA can also be used to verify the results of the more complex and therewith error-prone indirect force determination. If only the total structure-borne sound share of the tire-road noise or the contribution of the different axles and tires are of interest, the OTPA should be used because it is a quick method. Only four acceleration sensors and operational data are necessary. Elaborate impact hammer measurements must be performed if a more detailed path contribution is necessary.

An example of the results of indirect force determination is shown in Figure 6.10. The contribution shares are depicted as averaged spectra in a 2D-plot, each row represents one spectrum. Figure 6.10 top shows the contributions of the main parts of the front axle: transverse control arm, radius rod and strut bearing of the left and right vehicle side. As one example, in the bottom of Figure 6.10 all paths are depicted which contribute to the transverse control arm at the left side.

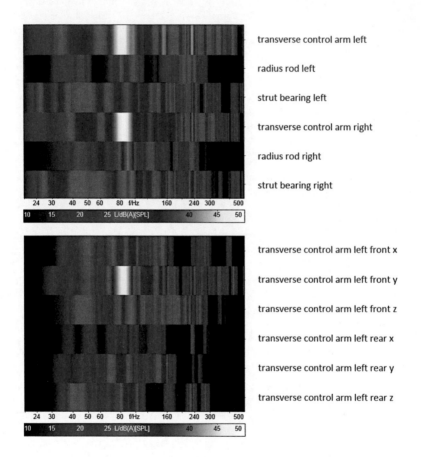

Figure 6.10: *Structure-borne sound shares of the tire-road noise on a medium rough road surface at a speed of 30 km/h using indirect force determination.*

6.3 Comparison of direct force measurement with indirect force determination

[3] A compact van is set up with four wheel force dynamometers of type RoaDyn System S625 2000 from Kistler [104] which are used in place of the standard wheels (Figure 6.11). In each case four load cells measure the force flow through a special rim in three spatial directions. The data is transmitted wirelessly via induction with a rotor and a stator. An on-board unit converts the forces of the rotating load cells into forces and moments related to the fixed vehicle coordinate system. The dynamometer rim weighs a bit more than the standard aluminum rim. In

Figure 6.11: *The wheel force dynamometer of type RoaDyn System S625 2000 from Kistler [104] is used in place of the standard wheel and allows for measurement of forces and moments in three dimensions.*

addition, a three-dimensional acceleration sensor is mounted on each wheel hub carrier. Further acceleration sensors are applied to the chassis, amongst others, on the transverse control arms, to allow an indirect force determination. On test tracks with different road surfaces, operational data were recorded covering different speed ranges.

On the test vehicle, impact measurements were performed in order to determine the wheel hub forces indirectly and compare them with the measured forces of the wheel dynamometer. The four wheel hubs in the three spatial directions are chosen as impact positions. Three-dimensional accelerometers were placed at these locations. Further sensors were applied to the structure to allow an over-determination.

The averaged spectra of the forces during a constant speed of 30 km/h on a medium rough road are shown in Figure 6.12. The influence of the engine noise on the forces

---

[3]This part has been previously published in [71].

is negligible. There is good accordance between the measured wheel forces (blue dotted curves) and the results of the indirect force determination (red curves). The shapes of the curves are almost identical. Nevertheless, at some frequencies the deviation is up to 10 dB. For other operating conditions like different road surfaces or vehicle speed these statements are also valid.

The indirect force determination is difficult to perform at frequencies below 60 Hz because during impact measurements the signal to noise ratio (SNR) of the acceleration signals could be low in this frequency range, leading to estimation errors in the inertances. This could be a reason for the higher deviations between measured and calculated forces below 60 Hz.

The accelerations on the wheel hub carrier can be caused by forces as well as moments induced by the wheel. The influence of the moments on the acceleration signals is present, but is neglected here. This could affect the calculated forces because the share of the moments in the acceleration signals is also related to the forces. Taking the moment excitations also into account by measuring angular accelerations and corresponding transfer functions could improve the indirect determination of the forces.

Furthermore, the assumption that the portion of the impact hammer's force acting on the wheel (source structure) is negligible must be proven. There are two ways to gain the operational forces instead of the blocked forces. First, the inertances may be measured while the source structure is separated from the receiver structure as mentioned above. However, dismounting the wheel without changing the system's behavior would be a challenging task. With dismounted wheel the hub carrier is in a different position, if the car is on a hydraulic lift. Then the static force distribution in the suspension system is different. Second, operational forces can be calculated from blocked forces using a correction function deduced from the determined mobility of the source structure as shown in [56].

In the measured as well as in the calculated forces symmetries between the left and right side of the vehicle are visible. For example, the force in x-direction of the front left and right wheel are very similar. Small deviations between the left and right side can be explained by the fact that the car is not completely symmetric, e.g., the gear box is located on the left side of the car. The differences between the front and rear wheels are obvious because of different suspension constructions. On the front axle there is a MacPherson strut and on the rear axle a multi-link suspension. Although the static wheel forces of the z-direction - the wheel loads - are much larger than the static values of x- or y-direction, between 20 Hz and 400 Hz for all directions the dynamic forces are within the range of -10 dB [N] and 30 dB [N].

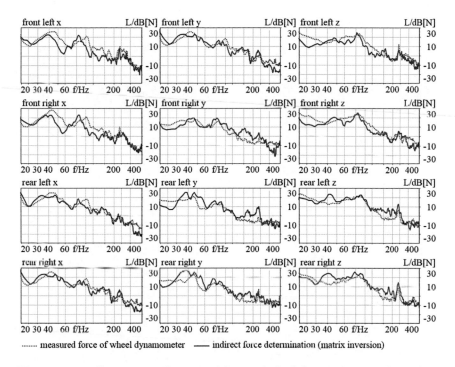

Figure 6.12: *Comparison of measured forces of wheel dynamometer with results of indirect force determination.*

Chapter 7

Conclusions and outlook

Transfer path analysis and synthesis (TPA / TPS) can be applied to find solutions for source-transmission-receiver issues during the development process of a vehicle. Which specific TPA method should be selected depends on the application. The selection is a compromise between quality of the results on the one hand and time and effort on the other hand. Operational transfer path analysis (OTPA) is especially suitable for tire-road noise analysis. It is a fast and easy-to-use method to extract contributions of the different tires or axle, as well as the structure-borne and airborne noise shares of the tire-road noise. One constraint is that the engine is switched off, e.g. during a coast-down, otherwise the tire-road noise synthesis would contain unwanted engine noise shares. But the engine or electric motor should be running if the tire-road noise is to be investigated under dynamic driving conditions. For example, this is the case if a driving cycle should be analyzed that is a more realistic representation of a customer's driving experience than a coast-down without a running engine. Another aspect is studying the influence of different drive torque on the tire-road noise depending on powertrain load, because changing driving forces may have an influence on the tire-road noise, especially in the case of sports cars. Here, the conventional approach cannot be used. That is the reason why an approach for the auralization of tire-road interior noise under dynamic driving conditions without disturbing noise shares from engine and wind flow is proposed. The key aspects of the new approach and other improvements of chassis TPA presented in this thesis are summarized in table 7.1.

With the presented crosstalk cancellation of the engine's influence the synthesis of tire-road interior noise is no longer limited to coast-down measurements. Now it is possible to auralize the tire-road noise while driving with constant speed or during a run-up without disturbing engine noise shares. The additional measurement effort is small. Only simultaneously measured accelerations on the engine (active-side) are needed. In a preprocessing step the influence of the engine on the tire input signals are eliminated. The key is that these active-side accelerations are not affected by tire-road noise because of soft engine mounts and high engine mass. These signals

Table 7.1: *Improvements of chassis TPA*

Auralization of tire-road noise under dynamic driving conditions

- separation of vehicle interior noise into the individual sound shares generated by engine, tire-road contact and wind flow
- based only on in-situ measured operational data, no time-consuming measurements of transfer functions
- small additional measurement effort (acceleration sensors placed on the engine)
- also suitable for hybrid and electric cars
- the synthesis of tire-road interior noise is no longer limited to coast-down measurements.

Investigations of chassis TPA

- OTPA cannot be used for a detailed partition of the tire-road noise into the contribution of single structure-borne transfer paths of the vehicle chassis. In this case an indirect force determination with matrix inversion should be performed.
- Results of OTPA can be used as an anchor to evaluate the reliability and quality of conventional TPA results.
- Wheel force dynamometers, which have been primarily developed to measure quasi-static forces for vehicle dynamics or durability tests, are also suited for NVH applications up to 400 Hz.

contain only engine noise, so it is possible to synthesize the engine noise shares of the tire-input signals in the time domain and to subtract them from the measurement. The result does not contain disturbing engine noise shares. A cancellation using subtraction can be performed, because the synthesis of the noticeable engine orders is precise enough in level and phase. The tire microphone signals are processed in the same way. The engine microphone signals cannot be used in the cross-talk cancellation, because they contain tire-road noise shares, too. It is utilized that there is a very high correlation between airborne sound radiation and engine vibrations. In the cross-talk cancellation only the correlation is evaluated so that the airborne sound radiation of the engine is not necessary for the crosstalk cancellation. In the investigated examples the correlation between engine airborne radiation and engine vibrations is sufficiently high. If the tire microphones would contain engine noise that is uncorrelated with the active-side engine accelerations, this approach is not well-suited. Then the method must be extended by a CTC of airborne sound. This requires further research. For example, in an iterative approach, microphone signals in the near field of the engine that also contain tire-road noise could be evaluated.

Electric cars have a different motor noise containing more high frequency components (e.g. they are caused by the converter) than orders at low frequencies like a combustion engine, but the proposed approach can be applied in their case too. Acceleration sensors for CTC should be placed on the electric motor and the housing of the converter. The positions of the sensors must be chosen with care, so that only vibrations of the electric powertrain are recorded and no tire-road excitations [98]. Otherwise the described cross-talk cancellation would not be possible.

For a more detailed partition of the tire-road noise into structure-borne transfer paths of the vehicle chassis OTPA cannot be used because the highly correlated signals would lead to unreliable or incorrect results. The inverse force identification is a suitable method with additional measurement effort. The operational forces are calculated indirectly from operational accelerations using the structure inertances derived from impact measurements.

Mounts are not considered in the indirect force determination method, but mounts play an important role in structure-borne transmission paths, especially for powertrain noise. They are used to decouple the source (engine) from the receiver structure (car body). The structure-borne sound can be varied in a wide range by changing mount parameters. Even close to the end of the development process, mounts can still be modified when it is too late for other countermeasures like increasing local stiffness. Mounts should be included in a transfer path model to discover more means of improving transmission behavior. This was the motivation to develop the effective mount transfer function method and to use parametric mount models. The improvements of powertrain TPA, proposed in this thesis, are summarized in table 7.2.

The effective mount transfer function method describes mount characteristics in the current assembly situation. It combines the advantages of both methods: inverse force identification with matrix inversion and mount attenuation function method. The coupling of the structure is considered and the auralization is performed by filtering active-side accelerations in the time domain. The quality of the synthesis is increased because active-side accelerations are less disturbed by other sources than passive-side accelerations, which are usually used in the inverse force identification method. The synthesized structure-borne engine shares based on active-side accelerations are much clearer and less noisy. If the coupling of the structure, which causes a crosstalk in the acceleration signals in the case of force excitations at several positions, were to be neglected, the operational forces would be overestimated, especially for low frequencies. The new method has been developed with a focus on the engine mounts of a car, but it is also suitable for other mounts or any machine that is elastically connected to a receiver structure [59].

The additional measurement effort compared to the standard inverse force identification with matrix inversion is marginal. Only some additional acceleration sensors placed on the engine are needed. The effective mount transfer functions are determined in-situ, that means, the engine or mounts must not be removed during the measurements. This saves time and prevents the system from being changed due to disassembling and reassembling. Measuring mount characteristics on a special test rig are omitted so that the challenges of crafting adaptors and setting the correct preload are avoided.

The effective mount transfer function method describes the mounts in the current situation. The contribution of the structure-borne sound paths can be calculated and critical paths can be identified. But this is only a first step, the second question is how to improve a complaint sound issue. Then mount models with physical parameters help to predict how to modify a mount. Softer mounts are often a preferable solution for a better NVH (noise, vibration and harshness) performance, but they must still be able to bear up the static load of the engine. The benefit of physical parameters like stiffness is that they allow for the possibility of checking if a desired modification is physically feasible. Parameter studies can be performed by simply varying these parameters virtually in the TPA model. The effect on the induced operational forces and thereby the contribution of the transfer paths to the interior noise can be evaluated. Another application is a sensitivity analysis, where the influence of parameter variations in the production process on the structure-borne sound is tested.

The effective mount transfer functions are a very good basis for estimating parameters of mount models. Simple models with only a few parameters have the

Table 7.2: *Improvements of powertrain TPA*

Effective mount transfer function method

- Mount characteristics are determined based on in-situ measurements.
- The additional measurement effort compared to the standard inverse force identification with matrix inversion is marginal.
- The coupling of the structure is considered and the synthesis quality is increased because of using active-side accelerations.

Parameter mount models

- Models with physical parameters help to predict how to modify a mount.
- Simple models with only a few parameters are more robust against overfitting than complicated models with many parameters.
- Kelvin-Voigt models already achieve good results. The parameters are estimated using in-situ measurements without special test rigs.
- The prediction quality can be increased if a directional coupling within a mount is considered.

Avoiding auralization artifacts

- An auralization of TPA results should not contain artifacts, which distract the listener from the proper sound, leading to another perception.
- For the widely used inverse force identification method with matrix inversion an approach is proposed that eliminates tonal components produced by the inversion of ill-conditioned matrices.
- Disturbing wrong peaks in the apparent mass transfer functions are detected and removed.

Transfer path synthesis based on engine test rig data

- Input signals of an engine measured on a test rig or at vehicle A can be used for a transfer path synthesis based on a TPA model of a different vehicle B, if the influence of the different load impedances is considered.
- Frequency dependent correction functions are determined from impact measurements of the assembled systems to adapt the acceleration signals.
- The feasibility of this method has been successfully demonstrated in an example with two miniature vehicle models.

advantage that they are more robust against overfitting than complicated models with many parameters. Overfitting means a very good prediction of the forces, but at the same time the parameters have no physical meaning any more. Kelvin-Voigt models already achieve good results. The parameters, stiffness and resilience, are estimated using in-situ measurements only without special test rigs. The prediction quality can be increased if a directional coupling within a mount is considered. That means an excitation in one direction also has an effect on the other directions. If three Kelvin-Voigt models per direction are used the estimated forces can be improved. The coupling of the structure is considered by using dense inertance matrices. The method has been successfully demonstrated on engine mounts and mounts of a triangle transverse control arm, which is an important part of the chassis.

The mount stiffness can vary with changing preload or excitation amplitude. The proposed method has the benefit of involving mount characteristics that are determined from an in-situ measurement under a specified load at a given well-defined operating condition. During this measurement excitation amplitude and preload is sufficiently constant in most cases. From this follows that mount stiffness changes are negligibly small. Of course, the Kelvin-Voigt model parameters and the calculated effective mount transfer functions are only valid for the operating condition that has been analyzed. For a measurement under varying load, a non-linear behavior of rubber mounts can be considered by using load-dependent effective mount transfer functions. For each load of interest, a set of transfer functions are calculated from the corresponding measurements. In the synthesis step the proper set of transfer functions is chosen according to the actual load.

It should be noted, that the synthesis quality of the new approaches depends on the quality of the indirect force determination. The indirectly determined forces are the basis for the calculation of the effective mount transfer functions. An over- or underestimation of the determined force cannot be detected because the effective mount transfer function is calculated using correlation analysis between engine vibration and indirectly determined force. The impact measurements, which are performed to determine structure inertances, are a critical point. They need most of the time of the data acquisition part, but they must be performed with utmost care to achieve the best possible results. Reaching a sufficient signal to noise ratio in all signals is a challenging task, especially for low frequency and distant sensors. Reproducible repetitions of single impacts are required but without damping the structure with the hammer tip. In a modern vehicle some positions can be difficult to access with an impact hammer or sometimes it is necessary to remove car components first, e.g., the starter battery. This could inadvertently change the system's behavior. Resulting consequences must be individually appraised in each case.

Kelvin-Voigt models are most suitable for conventional rubber-metal bearings. They can be applied to hydro mounts with restrictions if the mount characteristics can be linearized for the operating condition under test. Further research is necessary to develop and validate more complex models, for example, for switchable hydro-mounts. Also important is increasing the usable frequency range of the mount models, which could be relevant for noise issues regarding engine roughness. The effect of changes in load, preload or environmental conditions like temperature on the effective mount transfer functions and the resulting model parameters could be studied in detail. In a case where changing the parameter has no negligible influence, the mount model needs to be extended or the model and accordingly its parameters will only be valid for the current operating condition.

Binaural auralization is important for all applications. An auralization allows the engineer, a decision-maker or a customer to experience the results of a transfer path synthesis by listening. Judging the acoustic quality with your own ears is more re-liable and meaningful than looking at single values or diagrams. The auralization must not contain artifacts, which distract the listener from the proper sound, lead-ing to another perception. For the widely used inverse force identification method with matrix inversion an approach is proposed that eliminates tonal components produced by an inversion of an ill-conditioned matrix. These tonal components are generated by narrow peaks in the transfer functions. In terms of signal energy these artifacts are irrelevant. Nevertheless, they are able to attract the attention of the listener. The disturbing peaks in the apparent mass transfer functions are detected considering high values in the matrix condition and the wrong values are replaced by interpolation.

Binaural transfer path analysis and synthesis is a helpful methodology used in the development process of vehicles or other products and machines to identify the sound transmission from sources to receivers. The TPA/TPS methodology covers different transmission models and various approaches to determine the required transfer func-tions depending on the actual application. In this thesis enhancements have been proposed to analyze and auralize the sound transmission of the vehicle driving noise caused by powertrain, tire-road contact and wind flow. Future challenges of TPA include increasing the frequency range and prediction quality as well as reducing time and effort. Another point would be a better discrimination of the sound contri-bution caused by structure attachment points which are in very close proximity. The on-going progress made in simulation techniques should be utilized in the future by combining more simulated data with experimental data to use TPA at earlier stages in the development process.

Nomenclature

ω angular frequency

$\mathbf{a}$ acceleration vector

$\mathbf{a}^{AS}$ active side velocity vector

$\mathbf{a}^{PS}$ passive side velocity vector

$\mathbf{D}(f)$ diagonal inertance matrix

$\mathbf{F}$ force vector

$\mathbf{F}_L^A$ load force vector in case A

$\mathbf{F}_L^B$ load force vector in case B

$\mathbf{F}^{PS}$ passive side force vector

$\mathbf{H}_\Delta(f)$ correction function matrix

$\mathbf{I}$ inertance matrix

$\mathbf{I}^{MNT}$ mount inertance matrix

$\mathbf{M}$ apparent mass matrix

$\mathbf{v}_L^A(f)$ load velocity vector in case A

$\mathbf{v}_L^B(f)$ load velocity vector in case B

$\mathbf{v}^{PS}$ passive side velocity vector

$\mathbf{v}_S(f)$ source velocity vector in case B

$\mathbf{Y}_L^A(f)$ mobility matrix of receiver structure in the case A

$\mathbf{Y}_{SL}^{A}(f)$ mobility matrix of the assembled system composed of source structure S and receiver structure L in the case A

$\mathbf{Y}_{L}^{B}(f)$ mobility matrix of receiver structure in the case B

$\mathbf{Y}_{SL}^{B}(f)$ mobility matrix of the assembled system composed of source structure S and receiver structure L in the case B

$\mathbf{Y}_{S}(f)$ mobility matrix of source structure

$\mathbf{Z}^{PS,AS}$ mount transfer impedance matrix

$\mathbf{Z}^{PS,PS}$ mount output impedance matrix

$\tilde{F}_{k}^{IFD}$ force at position k calculated using IFD

$\tilde{F}_{k}^{MTF,model}$ force at position k calculated using effective mount transfer function from mount model

$\tilde{p}_{i}$ synthesized sound pressure at receiver position i

a acceleration

F force

F_{L}^{A} load force in case A

F_{L}^{B} load force in case B

F^{AS} active side force

F_{k}^{AS} active side force at position k

F^{PS} passive side force

F_{k}^{PS} passive side force at position k

F_{H} force of impact hammer

F_{L} load force

$G(f)$ transfer function

$G_{o}(f)$ averaged spectrum of time signal $s(t)$

$H(f)$ transfer function

$h(t)$ impulse response of $H(f)$

$h(t)_{k,i}^{ATF}$ impulse response of acoustic transfer function from force position k to receiver position i

$h(t)_{k,l}^{\mathbf{M}}$ impulse response from apparent mass matrices with index k and l

$H^{ATF}(f)$ transfer function with superscript ATF: acoustic transfer function

$H_{\Delta}(f)$ correction function

$H_k^{AMF}(f)$ transfer function with superscript AMF: apparent mass transfer function at position k

$H_k^{MAF}(f)$ transfer function with superscript MAF: mount attenuation transfer function at position k

$H_k^{MTF,model}(f)$ transfer function with superscript MTF: effective mount transfer function at position k based on a mount model

$H_k^{MTF}(f)$ transfer function with superscript MTF: effective mount transfer function at position k

$H_{i,j}(f)$ transfer function of a MIMO system from input with index j and output with index i

I (mechanical) inertance

i index of receiver positions

I^{MNT} mount inertance

I_k^{MNT} mount inertance with index k

$I_{k,k}^{STR}$ structure inertance with index k,k

$I_{k,l}$ inertance of force excitation point k and acceleration position l

j imaginary value

K number of force excitation points

k index of force excitation points

L number of acceleration positions

l index of acceleration positions

M apparent mass

$MIMO$ multiple input multiple output

n resilience

$n(t)$ noise time signal at the output of a MIMO system

p sound pressure

Q volume velocity

r viscous damping factor

$t(t)$ output time signal of a MIMO system without noise

v velocity

$v_L^A(f)$ load velocity in case A

$v_L^B(f)$ load velocity in case B

v^{AS} active side velocity

v_k^{AS} active side velocity at position k

v^{PS} passive side velocity

v_k^{PS} passive side velocity at position k

v_L load velocity

v_s source velocity

$X(f)$ fourier transform of $x(t)$

$x(t)$ input time signal of a MIMO system

Y mobility

$Y(f)$ fourier transform of $y(t)$

$y(t)$ output time signal of a MIMO system with noise

$Y_{SL}^A(f)$ mobility of the assembled system composed of source structure S and receiver structure L in the case A

$Y_{SL}^B(f)$ mobility of the assembled system composed of source structure S and receiver structure L in the case B

Y_l load mobility

Y_s source mobility

Z impedance

$Z_L^A(f)$ load impedance in case A

$Z_L^B(f)$ load impedance in case B

$Z^S(f)$ source impedance

Z_k^{KV} impedance based on Kelvin-voigt model at position k

Z^{MNT} mount impedance

Z^{STR} structure impedance

$Z_k(f)$ impedance at position k

$Z_l(f)$ load impedance

$Z_{11}, Z_{12}, Z_{21}, Z_{22}$ two-port impedances (z-parameters)

(E)MTF (effective) mount transfer function

AMF apparent mass function

AMTF apparent mass transfer function

AS   active-side

ATF acoustic transfer function

BTPA binaural transfer path analysis

BTPS binaural transfer path synthesis

CTC   crosstalk cancellation

FEA   finite element analysis

FRF   frequency response function

IFD   indirect force determination

MAF   mount attenuation function

MNT   mount

NVH   noise vibration harshness

OTPA operational transfer path synthesis

PS   passive-side

SDOF single degree of freedom

TPA   transfer path analysis

TPS   transfer path synthesis

TTCA triangle transverse control arm

Bibliography

[1]   Richard Lyon. *Designing for product sound quality*. CRC Press, 2000.

[2]   Karl Benz. "Fahrzeug mit Gasmotorenbetrieb". German. Pat. 37435. 1886.

[3]   David C. Quinn and Ruediger von Hofe. "Engineering Vehicle Sound Quality".
      In: *SAE Technical Paper 972063*. 1997. DOI: 10.4271/972063.

[4]   Peter E. Pfeffer and Stefan Sentpali. "Am Puls der Zeit - Fahren als Erlebnis".
      In: *ATZextra* 18 (2013), pp. 20–25.

[5]   Horst Klingenberg. *Automobil-Meßtechnik: Band A: Akustik*. Springer-Verlag
      Berlin Heidelberg, 1991.

[6]   Bernd Philippen, Roland Sottek, and Martin Spiertz. "Blinde Quellentren-
      nung von Fahrzeuginnengeräuschen". In: *Fortschritte der Akustik - DAGA
      2010*. 2010.

[7]   Philipp Sellerbeck and Christian Nettelbeck. "Door Operating Sound Im-
      provement based on Jury Testing and System Analysis". In: *CFA/DAGA'04*.
      Strasbourg, France, 2004.

[8]   Sebastian Merchel, Anna Lepping, and Ercan Altinsoy. "Hearing with your
      Body: The Influence of Whole-Body Vibrations on Loudness Perception".
      In: *The Sixteenth International Congress on Sound And Vibration, Krakow,
      Poland*. 2009.

[9]   *Vehicle Acoustics - Door Slam / Door Opening*. HEAD acoustics GmbH.
      Oct. 2015. URL: http://www.head-acoustics.de/eng/nvh_consulting_
      automotive_applications_door.htm.

[10] Peter Zeller. "Handbuch Fahrzeugakustik". In: Vieweg+Teubner, 2009. Chap. 1,
      pp. 1–3.

[11] Klaus Genuit. "A Special Calibratable Artificial-Head-Measurement-System
      for Subjective and Objective Classification of Noise". In: *Inter-Noise '86*.
      Cambridge, Massachusetts, USA, 1986, pp. 1313–1318.

[12] Hatmut Bathelt. "Analyse der Körperschall-Übertragungswege in Kraftfahrzeu-
      gen". In: *Automobil-Industrie* 1/81 (1981).

[13] Juha Plunt. "Finding and fixing vehicle NVH problems with transfer path
      analysis". In: *Sound and vibration* 39.11 (2005), pp. 12–17.

[14] *Installation and Operating Manual PCB Triaxial ICP Accelerometer 356A16.* PCB. 2008.

[15] Roland Kühn, Christoph Meier, and Horst Schulze. "Einsatz der TPA im NVH-Entwicklungsprozess Powertrain". In: *Automotive and Engine Technology: 5th Stuttgart International Symposium.* 2003. Chap. Akustik und Mechanik, pp. 47–58.

[16] Peter Genender et al. "NVH-Aspekte der Integration des Antriebsstrangs in das Fahrzeug". In: *MTZ - Motortechnische Zeitschrift* 63 (2002), pp. 470–477.

[17] Christian Nettelbeck and Klaus Genuit. "Binaural Transfer Path Analysis and Interior Noise Simulation for Vehicle Testbench Measurements". In: *The 32nd International Congress and Exposition on Noise Control Engineering.* Jeju International Convention Center, Seogwipo, Korea, 2003.

[18] Stefan Pischinger et al. "Schwingungen in Allradantrieben". In: *ATZ* 05 (2015).

[19] N. Moller and M. Batel. "Obtaining Maximum Value from Source/Path Contribution Analysis". In: *IMAC XXIV Conference & Exposition on Structural Dynamics.* 2006.

[20] Roland Sottek et al. "Description of Broadband Structure-borne and Airborne Noise Transmission from the Powerpoint". In: *FISITA 2006 World Automotive Congress.* JSAE, 2006.

[21] Roland Sottek and Bernd Philippen. "Synchronization of Source Signals for Transfer Path Analysis and Synthesis". In: *SAE Technical Paper 2014-01-2086.* 2014. DOI: doi:10.4271/2014-01-2086.

[22] J. S. Williams and G. C. Steyer. "Experimental Noise Path Analysis for Problem Identification in Automobiles". In: *IMAC XIII - 13th International Modal Analysis Conference.* 1995.

[23] Julius S. Bendat and Allan G. Piersol. *Random Data: Analysis and Measurement Procedures.* John Wiley & Sons, Inc., 1986.

[24] Kihong Shin and Joseph Hammond. *Fundamentals of Signal Processing for Sound and Vibration Engineers.* John Wiley & Sons, Inc., 2008.

[25] Bernd Philippen and Roland Sottek. "Bestimmung kritischer Übertragungspfade bei der BTPA". In: *Fortschritte der Akustik - DAGA 2015.* 2015.

[26] David Bogema. "Comparison of Time and Frequency Domain Source Path Contribution Analysis for Engine Noise Using a Noise and Vibration Engine Simulator". In: *SAE Technical Paper 2008-36-0509.* 2008.

[27] S. Allen Broughton and Kurt M. Bryan. *Discrete Fourier Analysis and Wavelets: Applications to Signal and Image Processing.* John Wiley & Sons, Inc., 2008.

[28] Michael Vorländer. *Auralization.* Springer, 2008.

[29] David Bogema. "Can You Hear It Now? Time-Domain Source-Path-Contribution Applied To a Diesel Engine". In: *SAE Technical Paper 2012-36-0626*. 2012.

[30] Jens Blauert and Klaus Genuit. "Evaluating sound environments with binaural technology-Some basic consideration." In: *Journal of the Acoustical Society of Japan* 14.3 (1993), pp. 139–145.

[31] Ercan M. Altinsoy. "Auditive Wahrnehmung und Beurteilung von instationären Fahrzeugaußengeräuschen". In: *Lärmbekämpfung, Zeitschrift für Akustik, Schallschutz und Schwingungstechnik* 9 (2014), pp. 64–71.

[32] Klaus Genuit and Wade Bray. "A Virtual Car: Prediction of Sound and Vibration in an Interactive Simulation Environment". In: *SAE Technical Paper 2001-01-1474*. 2001. DOI: 10.4271/2001-01-1474. URL: http://papers.sae.org/2001-01-1474/.

[33] Oliver Wolff and Roland Sottek. "Binaural Panel Noise Contribution Analysis - An Alternative to the Conventional Window Method". In: *CFA/DAGA '04, Strasbourg*. 2004.

[34] Oliver Wolff and Roland Sottek. "Panel Contribution Analysis - An Alternative Window Method". In: *SAE Technical Paper 2005-01-2274*. 2005.

[35] Andrea Grosso and Wilfred Hake. "Fast Panel Noise Contribution Analysis on a Running Train". In: *10ème Congrès Français d'Acoustique*. Lyon, France, 2010.

[36] Oliver Wolff, W. Tijs, and H-E de Bree. "In-flight panel noise contribution analysis on a helicopter cabin interior". In: *SAE Aerospace 2009, Seattle*. 2009.

[37] Roland Sottek. "Acoustical Relevance of Vibrating Structures". In: *Prooceedings of the IMAC-XXVII*. Orlando, Florida, USA, 2009.

[38] Juha Plunt. "Strategy for transfer path analysis (TPA) applied to vibroacoustic systems at medium and high frequencies". In: *ISMA 23*. Leuen, Belgium, 1998.

[39] Philipp Sellerbeck et al. "Improving Diesel Sound Quality on Engine Level and Vehicle Level - A Holistic Approach". In: *SAE Technical Paper 2007-01-2372*. 2007. DOI: 10.4271/2007-01-2372. URL: http://papers.sae.org/2007-01-2372/.

[40] Mark A. Gehringer. "Application of Experimental Transfer Path Analysis and Hybrid FRF-Based Substructuring Model to SUV Axle Noise". In: *SAE Technical Paper 2005-01-1833*. 2005.

[41] Krishna R.. Dubbaka, Frederick J. Zweng, and Shan U. Haq. "Application of Noise Path Target Setting Using the Technique of Transfer Path Analysis". In: *SAE Technical Paper 2003-01-1402*. 2003. DOI: doi:10.4271/2003-01-1402.

[42]   Andrei Cristian Stan et al. "Experimental Transfer Path Analysis for a Heavy Duty Truck". In: *Proceedings of the 12th Biennial Conference on Engineering Systems Design and Analysis - ESDA14*. Copenhagen, Denmark, 2014.

[43]   Carsten Zerbs and Ingmar Pascher. "Modifizierte TPA für die Prognose der Schallabstrahlung von Überwasser-Schiffen". In: *Fortschritte der Akustik - DAGA 2014*. 2014.

[44]   Stephan Schulze. "Betriebs-TPA zur Trennung von körper- und wasserschallinduzierten Eigenbootgeräuschen in Antennen". In: *Fortschitte der Akustik - DAGA 2014*. 2014.

[45]   "II.2 - Dynamic Response". In: *ISSC 2003 - 15th International Ship and Offshore Structures Congress*. Ed. by A. E Mansour and R. C. Ertekin. 2003, pp. 193–264.

[46]   Yi Wang et al. "Driveline NVH Modeling Applying a Multi-subsystem Spectral-based Substructuring Approach". In: *SAE Technical Paper 2005-01-2300*. 2005.

[47]   Frank J. Fahy and Paolo Gardonio. *Sound and structural vibration: radiation, transmission and response*. Academic press, 2007.

[48]   P. Gardonio and M. J. Brennan. *Mobility and Impedance Methods in Structural Dynamics: An Historical Review*. Tech. rep. 289. ISVR, Oct. 2000.

[49]   Frank J. Fahy. "The Vibro-Acoustic Reciprocity Principle and Applications to Noise Control". In: *Acta Acustica united with Acustica* 81.6 (Nov. 1995), pp. 544–558.

[50]   Roland Sottek, Philipp Sellerbeck, and Martin Klemenz. "An Artificial Head Which Speaks from Its Ears: Investigations on Reciprocal Transfer Path Analysis in Vehicles, Using a Binaural Sound Source". In: *SAE Technical Paper 2003-01-1635*. SAE 2003 Noise & Vibration Conference and Exhibition. 2003. DOI: doi:10.4271/2003-01-1635. URL: http://papers.sae.org/2003-01-1635/.

[51]   Udo Fingberg. "Statistical Noise Path Analysis". In: *IMAC XIV 14th International Modal Analysis Conference - Noise and Vibration Harshness (NVH)*. 1996.

[52]   *Data sheet Kistler 3-Component Force Link Type 9347C*. 2011.

[53]   Jürgen Bukovics et al. "Direkte Messung von Schwingungskräften - Ein Weg zu verbessertem Abrollkomfort". In: *ATZ Automobiltechinische Zeitschrift* 100 (1998), pp. 7–8.

[54]   Udo Fingberg and Thomas Ahlersmeyer. "Geräuschpfadanalyse einmal anders - Ein neuer Ansatz aus der Praxis". In: *3. Fahrzeugakustik. Tagung Geräuschminderung in Kraftfahrzeugen, Haus der Technik e. V.* 1992.

[55] Udo Fingberg. "Road/tyre noise development using noise path analysis techniques". In: *Proceedings of the 19th International Seminar on Modal Analysis.* Leuven, Belgium, 1994, pp. 899–910.

[56] A. T. Moorhouse, A. S. Elliot, and T. A. Evans. "In situ measurement of the blocked force of structure-borne sound sources". In: *Journal of Sound and Vibration* 325.4-5 (Sept. 2009), pp. 679–685.

[57] Bernd Philippen and Roland Sottek. "OTPA vs. TPA – comparison of Transfer Path Analysis methods". In: *Aachen Acoustics Colloquium 2011.* 2011.

[58] Bernd Philippen, Roland Sottek, and Payam Jahangir. "Comparison of directly and indirectly measured forces for tire-road noise analysis". In: *AIA-DAGA 2013 International Conference on Acoustics.* Merano, 2013.

[59] Roland Sottek and Bernd Philippen. "An Unusual Way to Improve TPA for Strongly-Coupled Systems". In: *SAE International Journal of Passenger Cars - Mechanical Systems* 6.2 (2013), pp. 1325–1333. DOI: doi:10.4271/2013-01-1970. URL: http://papers.sae.org/2013-01-1970/.

[60] Tom Knechten, Marius-Cristian Morariu, and PJG van der Linden. "Improved Method for FRF Acquisition for Vehicle Body NVH Analysis". In: *SAE Technical Paper 2015-01-2262.* 2015.

[61] Patrick Glibert and Nis Bjorn Moller. "Noise Path Analysis - A Tool for Reducing Testing Time". In: *IMAC XVII - 17th International Modal Analysis Conference.* 1999.

[62] E. H. Moore. "On the Reciprocal of the General Algebraic Matrix". In: *Bulletin of the American Mathematical Society* 26 (1920), pp. 394–395.

[63] Werner Biermayer et al. "Sound Engineering based on Source Contributions and Transfer Path Results". In: *JSAE Annual Congress.* Yokohama, Japan, 2007.

[64] H. G. Choi, A. N. Thite, and D. J. Thompson. "Inverse Force Determination: Refinements of Matrix Regularization and Sensor location Selection Methods". In: *ISVR Technical Memorandum* 924 (2003).

[65] A. N. Thite and D. J. Thompson. "The quantification of structure-borne transmission paths by inverse methods. Part 1: Improved singular value rejection methods". In: *Journal of Sound and Vibration* 264 (2003), pp. 441–431.

[66] A. N. Thite and D. J. Thompson. "The quantification of structure-borne transmission paths by inverse methods. Part 2: Use of regularization techniques". In: *Journal of Sound and Vibration* 264 (2003), pp. 433–451.

[67] Andre Elliot. "characterisation of structure borne sound sources in-situ". PhD thesis. University of Salford, 2009.

[68]   Andre Elliot and A. T. Moorhouse. "characterisation of structure borne sound sources from measurements in-situ". In: *Acoustics '08*. Paris, 2008.

[69]   N. Zafeiropoulos et al. "A Comparison of two In-situ Transfer Path Analysis Methods". In: *RASD 2013 11th, International Conference on Recent Advances in Structural Dynamics*. Pisa, 2013.

[70]   A.S. Elliot et al. "In-situ source path contribution analysis of structure borne road noise". In: *Journal of Sound and Vibration* 332 (2013), pp. 6276–6295.

[71]   Bernd Philippen et al. "Tire-road noise analysis using wheel force dynamometers". In: *Aachen Acoustics Colloquium 2012*. 2012.

[72]   Heinz-E. Meier and Bernd Wunsch. "Fahrzeuglager als mitbestimmendes Element der Geräuscheinleitung". In: *Elastische Lagerungen im Automobilbau, Haus der Technik*. Essen, 1995.

[73]   Elmar David, Rolf Helber, and Helmut Rees. "Methode zur Ermittlung der Übertragungssteifigkeit von Gummimetallteilen im akustischen Frequenzbereich". In: *ATZ Automobiltechinische Zeitschrift* 89 (1987), pp. 315–317.

[74]   C. Molloy. "Use of Four-Pole Parameters in Vibration Calculations". In: *The Journal of the Acoustical Society of America (JASA)* 29.7 (1957), pp. 842–853.

[75]   M. Krämer and R. Helber. "Methode zur quantitativen Analyse der Körperschallübertragung in Kraftfahrzeugen". In: *VDI-Fortschrittberichte Reihe 12 - 14. Internationales Wiener Motorensymposium* 182 (1993), pp. 362–379.

[76]   Roland Sottek and Bernhard Müller-Held. "NVH tools and methods for sound design of vehicles". In: *Inter-noise 2007*. Istanbul, Turkey, 2007.

[77]   E. Seidel. "Wirksamkeit von Konstruktionen zur Schwingungs- und Körperschalldämmung in Maschinen und Geräten". In: *Schriftenreihe der Bundesanstalt für Arbeitsschutz und Arbeitsmedizin*. Forschung Fb 852. 1999.

[78]   D. Göhlich and B. Köder. "VDI / FVT Jahrbuch 1993 Fahrzeug- und Verkehrstechnik". In: 1993. Chap. Berechnung der Körperschallübertragung bei Elastomerlagern, pp. 213–235.

[79]   J. W. Verheij. "Multi-path sound transfer from resiliently mounted shipboard machinery". PhD thesis. TU Delft, 1982.

[80]   John Dickens. *Investigation of Asymmetrical Vibration Isolators for Maritime Machinery Applications*. Tech. rep. DSTO Aeronautical and Maritime Research Laboratory, 1999.

[81]   P. Perry Gu and Joe Juan. "Application of Noise Path Analysis Technique to Transient Excitation". In: *SAE Technical Paper 972034*. 1997. DOI: 10. 4271/972034.

[82] Hendrik Sell. "Charakterisierung des dynamischen Verhaltens von elastischen Bauteilen im Einbauzustand". PhD thesis. TU Hamburg-Haburg, 2004.

[83] Thomas Hansen. "Analysis of Elastomer Components for Powertrain and Chassis Suspension". In: *ANSYS Conference & 26th CADFEM Users' Meeting*. 2008.

[84] Klaus Genuit and J. Poggenburg. "The Design of Vehicle Interior Noise Using Binaural Transfer Path Analysis". In: *SAE Technical Paper 1999-01-1808*. 1999. DOI: doi:10.4271/1999-01-1808. URL: http://papers.sae.org/1999-01-1808/.

[85] X. Bohineust et al. *AQUSTA - Improvement of the structural Acoustic Quality of transportation vehicles Using Simulation Techniques of binaural Analysis*. Syntheis Report for Publication. Synthesis Report BRE2-CT92-0238, June 1996. URL: http://cordis.europa.eu/documents/documentlibrary/26840171EN6.pdf.

[86] Klaus Genuit. "Application of Binaural Transfer Path Analysis to Sound Quality Tasks". In: *IMechE*. 2000.

[87] Roland Sottek and Bernd Philippen. "TPA innovations for strongly coupled systems". In: *Aachen Acoustics Colloquium 2013*. 2013.

[88] Kousuke Noumura and Junji Yoshida. "Method of transfer path analysis for vehicle interior sound with no excitation experiment". In: *FISITA 2006 World Automotive Congress*. 2006.

[89] Fabio Bianciardi, Karl Janssens, and Laurent Britte. "Critical Assessment of OPA: Effect of Coherent Path Inputs and SVD Truncation". In: *ICSV20 - 20th International Congress on Sound & Vibration*. 2013.

[90] Peter Gajdatsy et al. "Critical assesment of Operational Path Analysis: mathematical problems of transmissibility estimation". In: *Acoustics 08 Paris*. 2008.

[91] Peter Gajdatsy et al. "Application of the Transmissibility Concept in Transfer Path Analysis". In: *Mechanical Systems and Signal Processing* 24.7 (2010). Special Issue: ISMA 2010, pp. 1963–1976. DOI: 10.1016/j.ymssp.2010.05.008. URL: http://www.sciencedirect.com/science/article/pii/S0888327010001512.

[92] Bernd Philippen. "Operational Transfer Path Analysis mit Randbedingungen". In: *Fortschritte der Akustik - DAGA 2014*. 2014.

[93] Roland Sottek and Bernd Philippen. "Tire-Road Noise Analysis of On-Road Measurements under Dynamic Driving Conditions". In: *SAE International Journal of Passenger Cars - Mechanical Systems* 5.3 (2012). 7th International Styrian Noise, Vibration & Harshness Congress: The European Automotive Noise Conference, pp. 1116–1123. DOI: doi:10.4271/2012-01-1550. URL: http://papers.sae.org/2012-01-1550/.

[94] Karl Janssens et al. "OPAX: A new transfer path analysis method based on parametric load models". In: *Mechanical Systems and Signal Processing* 25 (2011), pp. 1321–1338.

[95] Bernd Philippen. *Untersuchungen zur Quellentrennung von Fahrzeuginnengeräuschen.* diploma thesis, RWTH Aachen University. 2008.

[96] Dietmar Gross, Wernder Hauger, and Peter Wriggers. *Technische Mechanik: Band 4: Hydromechanik, Elemente der Höheren Mechanik, Numerische Methoden.* Springer-Verlag, 2007.

[97] Bernd Philippen and Roland Sottek. "Parameterizing mount models from in-situ measurements". In: *SAE Technical Paper 2015.* 2015-01-2280.

[98] Roland Sottek and Bernd Philippen. "Advanced Methods for the Auralization of Vehicle Interior Tire-Road Noise". In: *SAE Technical Paper 2012-36-0640.* SAE Brasil NVH. 2012. DOI: 10.4271/2012-36-0640. URL: http://papers.sae.org/2012-36-0640/.

[99] Roland Sottek and Bernd Philippen. "Characterizing Tire and Wind Noise using Operational Path Analysis". In: *NAG/DAGA 2009, International Conference on Acoustics.* Rotterdam, 2009.

[100] Dennis de Klerk and Alexander Ossipov. "Operational Transfer Path Analysis: Theory, Guidelines & Tire Noise Application". In: *Mechanical Systems and Signal Processing* 24.7 (Oct. 2010). ISMA 2010, pp. 1950–1962. URL: https://www.isma-isaac.be/past/conf/isma2010/proceedings/papers/isma2010_0021.pdf.

[101] Christian Nettelbeck, Daniel Riemann, and Philipp Sellerbeck. "Road Noise Analysis Using A Binaural Time Domain Approach". In: *CFA/DAGA'04.* 2004.

[102] Norbert Wiener. *Extrapolation, Interpolation and Smoothing of Stationary Time Series.* Wiley, 1994.

[103] Roland Sottek and Bernd Philippen. "Separation of Airborne and Structure-Borne Tire-Road Noise Based on Vehicle Interior Noise Measurements". In: *SAE Technical Paper 2010-01-1430.* 6th International Styrian Noise, Vibration & Harshness Congress - Sustainable NVH solutions inspired by ecology and economy. 2010. DOI: 10.4271/2010-01-1430. URL: http://papers.sae.org/2010-01-1430/.

[104]   *Data sheet Kistler RoaDyn S625 System 2000*. URL: www.kistler.de.

Curriculum Vitae

Personal Data

04.03.1983	Bernd Philippen born in Heinsberg, Germany

Education

1993 – 2002	Kreisgymnasium Heinsberg

Higher Education

10/2003 - 12/2008	RWTH Aachen - Computer Engineering (Dipl.-Ing.)

Professional Experience

01/2016 - today	HEAD acoustics GmbH Product Manager
02/2009 - 12/2015	HEAD acoustics GmbH Research NVH - Project Engineer
10/2007 03/2008	HEAD acoustics GmbH Internship
01/2005 - 09/2007	Uniklinik RWTH Aachen Institut for Medical Psychology und Medical Sociology student assistent
05/2003 - 08/2003	Seniorenzentrum Breberen temporary employee

Heinsberg, Germany, Mai 12, 2016

Bisher erschienene Bände der Reihe

Aachener Beiträge zur Technischen Akustik

ISSN 1866-3052

Alle erschienenen Bücher können unter der angegebenen ISBN-Nummer direkt online
(http://www.logos-verlag.de) oder per Fax (030 - 42 85 10 92) beim Logos Verlag
Berlin bestellt werden.